The Handbook for the Apprentice of Biomedical Research

THE TOOLS OF SCIENCE

The Handbook for the Apprentice of Biomedical Research

THE TOOLS OF SCIENCE

Jose Russo

Fox Chase Cancer Center, USA

World Scientific

NEW JERSEY · LONDON · SINGAPORE · BEIJING · SHANGHAI · HONG KONG · TAIPEI · CHENNAI

Published by

World Scientific Publishing Co. Pte. Ltd.

5 Toh Tuck Link, Singapore 596224

USA office: 27 Warren Street, Suite 401-402, Hackensack, NJ 07601

UK office: 57 Shelton Street, Covent Garden, London WC2H 9HE

British Library Cataloguing-in-Publication Data
A catalogue record for this book is available from the British Library.

THE TOOLS OF SCIENCE
The Handbook for the Apprentice of Biomedical Research

ISBN-13 978-981-4293-16-7
ISBN-10 981-4293-16-4

Typeset by Stallion Press
Email: enquiries@stallionpress.com

Printed by FuIsland Offset Printing (S) Pte Ltd. Singapore

Contents

Acknowledgment

I dedicate this book to all my trainees from whom I have received more than they ever have expected to give me. Among them Jwang Ling MD, Eugene Agnone MD, Francisco Martinez MD, Gloria Calaf PhD, Daniel R. Ciocca MD, Lee K. Tay PhD, Stephen P. Ethier PhD, Gustavo A. Moviglia MD, Satori Higa MD, Megan Mills PhD, Gustavo Rubio Coronel MD, Anna Sapino MD, Fulvio Basolo MD, Gabriella Fontanini MD, Josiah Ochieng PhD, Muneesh Tewari MD, PhD, Pei-Li Zhang PhD, Maria Elena Alvarado MD, Anthony Magliocco MD, Teh-Yuan Ho PhD, Nandita Sohi PhD, Chai Yu-Li MD, Saad El-Gendy PhD, Judith Gordon MD, Eric G. Thomas DO, Roupen Yagsessian MD, Kunle Adesina MD, PhD, Ruth Padmore MD, PhD, Hu Yun Fu PhD, Ana Maria Salicione PhD, Betsy Bove PhD, Yajue Huang PhD, Ismael Dale Cotrin Silva, MD, PhD, Abdel-Rahman N. Zekri MSc, PhD, Xiaoshang Jiang PhD, Raquel Angela Silva Soares Lino PhD, Hasan M. Lareef MD, Gabriela Balogh, PhD, Fathima Sheriff MD, Sandra Fernandez PhD, Johana Vanegas MD, Raquel Moral PhD, Daniel Mailo PhD, Ricardo Lopez PhD, Julia Pereira PhD, Hilal I. Kocdor MD, PhD, and Mehmet A. Kocdor MD, Slyvana Carrea de Noronha PhD and Samuel Ribeiro de Noronha PhDc.

My special acknowledgment and thanks to Ms. Patricia A. Russo for her insightful editorial and stylish suggestions.

Thanks to Dr. Irma H. Russo for her critiques and delightful moments discussing the manuscript and ideas, to Ms. Jennifer McDonald for verifying the accuracy of the references and the search for not easy to find articles and the Pathology Consultation Services from Rydal, PA, that have financed the writing and editing of this book.

Jose Russo, MD

Chapter 1

Profile of a Scientific Researcher
(*Per scientiam ad sapientiam*)

Introduction

I have enjoyed science and everything related to medicine for as long as I can remember. From very early on I knew that these interests were more than passing curiosities, rather deeply rooted in my sense of vocation, even before I had a name for it or fully understood the meaning of this calling. Over the years I've asked myself what it was about this particular vocation that called me so strongly. Several possibilities come to mind, yet none of them were obvious to me at first. I do not come from a family of researchers; my father was a carpenter and my mother a homemaker, and while nurturing and supportive, they most certainly did not point me in the direction of the sciences, or any other direction for that matter. Their main interest was that I grow up to be an honest, and hardworking individual like them and all my Italian ancestors. They did, however, notice how science excited me and supported my dreams and aspirations in whatever ways they could. This is perhaps one of the most important lessons I have learned from my parents — that we must respect and nurture our children's dreams because we never know where they might lead them. With my parent's teaching as an example, I have tried to do exactly that with my own child.

As a child I was fascinated by glass bottles and had amassed quite a varied collection. In my mind, these were the makings of a chemistry laboratory, each one holding a different imaginary chemical or solution. For my eighth birthday my parents gave me a real chemistry set and the experiments that had once been relegated to fantasy could actually be performed.

I set up my "lab" in a small area of our family home where I collected insects, bones and other mysterious natural objects and played surgeon with my sister's dolls. It didn't take me long to realize that there wasn't much to be learned from pulling apart thread and stuffing, so from then on I dissected every single dead animal I could get my hands on. I called that space my "chemistry cabinet," imagining a set-up like the one assembled by Grand Duke Peter Leopold in the late eighteenth century so he could pursue his scientific passions. A major addition to my cabinet was the acquisition of a monocular microscope with 20× magnification that I purchased at Lutz Ferrando, an optical store in Mendoza, with money that I saved from helping my father with his carpentry work. My childhood friend Francisco Fisichella and I spent our summers and most of our free time dissecting frogs and rodents and obsessively pursuing the task of making those tissues grow *in vitro*, unsuccessfully. It was around that time that I learned the meaning of the word "cancer," which in my rudimentary understanding, meant an abnormal growth of cells that could kill a person. In those days the Internet did not exist so the public library was the only source of knowledge. It was in searching those great dusty science books, apparently not appealing to the general public since I was often the only one there, that my intellectual horizons started to broaden and the desire to be a physician-scientist was crystallized in my mind. At that point, just a few months after my eleventh birthday, it became clear to me that studying medicine was my only option. This was an important realization because I then had to choose the "college" that would enable me to do cancer research.

My five years at the Agustin Alvarez National College in Mendoza were fascinating training years for both my intellect and will. I woke up at five in the morning and studied six to seven consecutive hours before my classes started at four in the afternoon, not ending until eight. I learned to take brief five-minute naps every hour, to read out loud while standing to avoid falling sleep. This exercise was so good that when I needed to study for 12- to 14-hour stretches in medical school it wasn't difficult for me. I had excellent teachers and they taught me more than they were expected to and helped me gain access to every chemistry or physics laboratory available in my town, as well as the best collections of biological specimens available to us in the region. I finished with honors, an achievement I have come to appreciate more in retrospect. My goal was to gain admission to a medical school which accepted only 80 students with the highest scores on a series of dreadful written and oral examinations in English, Chemistry, Physics, Biology and Mathematics. I succeeded having the highest score of all the applicants. This moment still brings me warm memories and a feeling of satisfaction, not because I did so "quantifiably" well, but because the day I learned of my admission to medical school was the first day of my long journey as an *Apprentice of Science*.

In reflecting how my own scientific research vocation has emerged, and in watching students struggle with their own questions of vocational identity, I have come to realize that there is a need to outline and analyze how this inner calling is recognized. By examining my own experience, as well as those of other scientists who have paved the road before me, it is my hope that the new generation of budding scientists will be inspired *to listen closely to the forces within them to discover their own special calling*. The best piece of advice I can give to anyone, not just those interested in science, is to find out as early as possible what you want, and train yourself to be resilient and focused. These are the two most important qualities one must have to reach their goals and live their dreams. But as important as this advice could be, it is also a mentor's function to recognize the inner calling of future scientific apprentices, by seeing beyond appearances to the true scientific apprentice within. I would like to emphasize especially the point of *learning to see beyond appearances*. This means that the mentor must see the inner person and look past any gender, ethnic or physical bias. The mentor must meet the mind and soul of the scientific apprentice by listening to them, and only then will the real scientific apprentice make him or herself known. Like a good parent, the mentor should support the apprentice in any way they can.

I have trained more than 50 PhDs and MDs in cancer research and I have witnessed how each one of them, at different times, have come to this internal struggle of whether they want to be an Apprentice of Science or not. For each of them the call was different, but their common path was that at their turning point they all discovered what was needed to be a scientific apprentice. There have also been a similar number of high school and college students that have been interns for different periods of time in my laboratory and each of them has a special place in my thoughts. I remember one of them in particular, Muneesh Tewari. I met him when he was 14 years old, just a junior in high school. Nobody wanted to take him as an intern in their laboratory because he was too young and no one believed he was serious about cancer research. This was in 1986, and in those days high school students were not easily accepted into working laboratories. When I met him I realized that he had a special inner drive and I did not have any doubt that he was a scientific apprentice at heart. We needed to obtain permission from our institution at that time, the Michigan Cancer Foundation, to allow him to be in the laboratory. But once we did, Muneesh spent two years with us working after school. In that time he coauthored one publication with us[1] and obtained more scholarships than he could use based on his work with our group. He graduated in 1990 Summa Cum Laude from Case Western Reserve University. After college, he opted for a PhD and MD program at the University of Michigan and graduated in 1997. Dr. Tewari is now an established Research Scientist who has made important contributions regarding the role of microRNA in cancer.

Profile of the Research Scientist

All too often the term "scientific researcher" elicits a picture of *Back to the Future's* Dr. Emmett Brown, eyes bulging with excitement, a shock of white hair poking through his latest contraption. Popular culture tends to view the scientist as "mad," blindly driven, socially obtuse, and completely lost in their own universe of inquiry. While the best scientists are driven and deeply engaged in their field of study, the typical profile of a researcher, as portrayed by film and television, is misleading. But it's not a researcher's physical assets that are of real interest here, rather, the inner self that emerges when an individual uses his or her personal talents to ask specific scientific questions.

Determining when an individual goes from being simply inquisitive to becoming a true researcher at heart is not an easy task. Often there is a trigger — an event or some other kind of catalyst — that pushes a person into this other realm of scientific exploration. For Robert Koch, who in 1877 isolated the tuberculosis bacillus and is considered the father of bacteriology, it was probably when his young wife bought him a microscope for his 30th birthday. His wife's support, as much as the gift itself, greatly encouraged him. As a physician, Koch had grown bored in his medical practice, something his wife might have sensed. Her sensitivity to her husband's unhappiness motivated her to give the gift that proved to be a turning point in Koch's career. What we now consider to be a common laboratory tool opened Koch's eyes to an unknown world. While the microscope itself didn't make him a scientist or give him an inquisitive mind, it did provide the means for him to answer deeper questions that up until that point had laid dormant in his mind. The introduction of a microscope into Koch's life was probably the trigger he needed in order to begin to know which questions to ask. By breaking away from the limitations of what can be seen by the naked eye, Koch entered a richer, more complex level of biological understanding. The support of loved ones such as his wife, are not to be discounted; her support may have indeed played a significant role in his motivations. Yet these are external parameters of an internal change that shaped the direction of Robert Koch's life. This internal catharsis, as real and forceful as it is, has seldom been written about.

The ability to ask specific questions is the basis of scientific inquiry and a fundamental characteristic of a research scientist. In the case of Francis Collins, mathematics captured his attention at an early age, but it was his study of medicine, mainly in the area of pediatrics, that paved the way to understanding the genomic basis of childhood diseases, which ultimately led to one of the most important scientific projects of the 20th century: the Human Genomic Project. Other researchers, like Gregory Mendel, held more humble aspirations. Mendel's dream was to be a certified teacher in a secondary school, and while he never passed the teaching exam, his perseverant work with peas led him to discover the basics of genetics and establish the laws that carry on his name. He died without knowing that his work would be seminal in establishing the basis of modern genetics.

Individuals like Louis Pasteur were scientifically and mentally prepared to make discoveries. Pasteur's greatest discoveries lie in molecular asymmetry, fermentation, cholera, spontaneous generation, rabies and many other industrial problems that allowed him to establish the basis of modern chemistry and microbiology. These findings originated from practical problems that needed solutions and he worked tirelessly until these challenges were met. Does the fact that his greatest intellectual achievements were derived not from the whim of inspiration but from real-life situations diminish his work? Not at all. Pasteur developed the scientific apprentice within by readily responding to challenges posed. We might ask ourselves whether his accomplishments would be the same if he had been the one to pose the original questions, to extract theoretical inquiry from his own mind; but in the end it doesn't matter. The thing to remember is that the way into scientific apprenticeship is multi-channeled. Not everyone knows what excites them on their own; sometimes it takes an unplanned problem to uncover that desire for inquiry that resides within. Historical timing, as much as personal timing, as well as the social environment, could make all the difference. This is why today we consider Darwin, and not his contemporary, Alfred Russel Wallace, the father of the evolution theory.

It is evident that Santiago Ramón y Cajal's father had a deep influence in his life and that he regarded his progenitor as a central force in the development of his character. In his book *Infancia y Juventud* which was published in 1920,[2] Cajal wrote, "*[my] father had this sacred anguish for poverty that had left in him a hard heart for fighting against misery injustice and despair.*" This philosophy of life was probably what eventually triggered in Cajal the need to stabilize and find a more coherent focus in his life. Before he began studying medicine under his father's suggestion, he had dabbled in various kinds of work, including painting, gymnastics — there was even a period in which he worked for a barber. In 1868, at the age of 18, Cajal's father sent him to study osteology in a barn adjacent to their home. Once settled into the practice of medicine, it is possible that he realized what he could do for others through his practice was limited. It's possible that this realization awakened a need in him for permanent solutions — or truth — much like that sought by research scientists. As rewarding as the practice of medicine is, the pain of being unable to solve the vast range of problems that take the life of so many can make us feel the despair of our limitations. It was his beginning in osteology that taught Cajal to perceive and analyze small details, like those in the apparently inert and irrelevant surface of bones. Over time Cajal recognized that his father was a genuine teacher who gave him a glimpse of what turned out to be the scientific apprentice within. Is this late blooming a bad sign? Not at all, but it is important to recognize the call when we hear it. Cajal was a late bloomer but his work on the central nervous system, neuroanatomy and descriptions of neural structures have vastly shaped our contemporary understanding of the human body.

In terms of how a scientific apprentice's mind works, all we have are clues. An adolescent Mendel foresees his destiny in a poem* written while studying at the gymnasium:

> May the might of destiny grant me
> The supreme ecstasy of earthly joy,
> The highest goal of earthly ecstasy,
> That of seeing, when I arise from the tomb,
> My art thriving peacefully
> Among those who are to come after me.

Besides this sense of having a destiny, we do not know if becoming a scientist was a burning goal for Mendel, or whether his achievements were the consequence of how he happened to be oriented by his professors. From elementary school he went to gymnasium, then on to the Institute of Philosophy, followed by entry into the monastery. It was there that Abbot Napp gave him the task of studying the generation of hybrids using *Pisum sativum*, otherwise known as the snow pea, which led to the discovery of what we now refer to as Mendelian genetics. This once again brings us to

---

*Robin Marantz Menig. *The Monk in the Garden: The Lost and Found Genius of Gregor Mendel, the Father of Genetics*, pp. 17–18, 2000.

the question of how mentoring shapes a scientist's formative ideas. That Mendel and Cajal's subject matter were significantly influenced by their mentors does not lessen the uniqueness of their work. It was the process of mentorship that brought out the talent of these two individuals and nurtured it to its full potential.

It was to a certain extent because of the monastic life at St. Thomas that Mendel was able to develop the discipline and methodic approach required to carry out his work. His days were organized for him. In his autobiography Cajal recounts how he needed to organize his days and clearly laments his formerly disorganized ways. For Cajal, it was a struggle to take hold of his intellectual life, but he succeeded with hard work and discipline.

Each individual is unique to his or her period of time and there is no particular pattern from which a scientist emerges, at least in the initial stage of their lives. Therefore we must ask ourselves: how can we recognize the budding scientist in their first manifestations? In an era in which science is extremely competitive and requires a significant amount of training, the earlier the individual finds out that science is part of their life endeavor, the better equipped they will be for their personal future and the future of science in their country and around the world.

Curiosity and Urge for Rational Understanding

Certain traits such as a heightened level of curiosity can sometimes be a sign that a child might be destined for a life in science. Certain people are hard-wired to ask questions about how and why things are the way they are. This innate need to understand the world's workings is what separates a person with an inclination for science from others. While people of all ilk and disciplines are *curious*, scientific curiosity is different in that it comes with an absolute need for understanding, a hunger for truth. This hunger is one of the basic and primitive properties that indicates the budding of a future scientist. When looking for scientific potential, it is important to consider the ability of this budding scientist to observe, see and pay attention to things and events that other individuals might not. A true scientist must be able to see beyond the apparent. The ability to listen and capture concepts or communicate and articulate what they are learning is also another property that should be observed in those who are considering the path of scientific research. This is not to say that there are fixed rules about who will and who will not become a scientist. On some occasions an individual possesses all these properties yet does not discover them until later on in life. Searching for these early signs is an important duty for teachers, parents and others in influential positions — these are the figures who play an important role in identifying and helping to flesh out potential, be it in research or any other field.

How do we know when someone has what it takes to embark on the journey of scientific apprenticeship? There is no fixed set of rules, but the ability to analyze a problem and persist until it is solved is a good beginning. Many major discoveries are made not due to special talents, but to common sense and hardwork — two of the most important ingredients. Some people have great memories, learn quickly, are mentally dexterous and skilful at communicating. Without a doubt these people, if perseverant and focused in their scientific endeavor, can make great contributions. However, most individuals are not endowed with *all* of these qualities, yet still make great contributions to science by using their talents wisely and forcing themselves to succeed. The key is having curiosity, a genuine desire to know, and not being afraid to ask questions. Curiosity, the ability to think logically, and rigorous work habits are the three main components that will define a scientific apprentice's talents.

Whether or not a scientific apprentice has the ability to think logically and independently is easy to detect. Those who consider the written word the final word, who see no need to reexamine or reinterpret a published paper, probably do not grasp the true definition of "research," which is *a diligent and systematic inquiry or investigation into a subject in order to discover or revise facts.* The word derives from a French word, *rechercher*, meaning "to seek." This entails challenging a hypothesis and going even further to consider that there may be other explanations for the same hypothesis. To see a lag in the knowledge, to understand that what is known in a biological process does not cover all angles, and that many questions remain unanswered, is the beginning of the scientific process, and as will be discussed in the next chapter, this is also the way to find one's research idea, or theme.

The scientific apprentice must also understand that an inquisitive sense of perception is of great importance for self discovery, but it is the fitness to persevere with an idea and concentrate on a specific problem that counts in the end. It is similar to a marathoner who may run for the pure joy of it, but knows it is the will to train and stay focused on the race that counts towards reaching the finish line. Just like the endorphins that fuel the marathoner for 26.2 miles, the scientific apprentice needs the same kind of stamina to carry through a new concept, idea or hypothesis. To acquire this training is not easy but once "muscle memory" is established, it becomes an integral part of the scientific apprentice's life. The reality is that an idea or problem changes from beginning to end and it is the continuous challenge of solving a constantly evolving question that shapes our neural network. This process requires solitude and on some occasions total isolation from noise and other external distractions. This self-imposed retreat is part of learning the intellectual introspection, which is key to the scientific apprentice's success. The scientific apprentice must learn not to be derailed by disturbing news and to stay away from the daily gossip that takes place in any academic institution. Keeping the mind focused during this period of mental training is crucial to succeeding and must become second nature. If the scientific apprentice cannot accomplish or believe in the importance of this type of "fitness," it is probably time to reconsider science as a way of life.

In the end, the great triumphs of science have resulted from the determination to follow an idea, as with Marie Curie's discovery of radium, or Newton's discovery of gravitational law. Edison put it well when he wrote that "discovery is ten percent inspiration and ninety percent perspiration." The scientific apprentice must remember that like the marathoner, crossing the finish line of an idea is the result of tremendous effort and pain that are overcome only by training and fitness.

The scientific apprentice must fight against the inertia of those who are blind to the adventure of knowledge, not impressed by anything and blindly accept natural phenomena without wondering about its nature. This inertia, and as I will discuss later on, the cynicism of the present day, has delayed our human advances as well as caused the discouragement of many good scientific apprentices.

It is worth mentioning some examples that illustrate how inner motivation and historical conditions contribute to the formation of the scientific apprentice, and how the work of one may have an expansive influence on so many others. Don Fawcett was born in 1917 on a farm in Iowa and attended Harvard University in 1934 where he "sampled the humanities," but soon he became fascinated with biology. Although his initial aspiration was surgery, he found that academia was his main inclination and ended up joining the Department of Anatomy at Harvard as a junior faculty member. At the same time Keith Porter had just published the first electron micrographs of tissue culture cells and the Department of Anatomy at Harvard had acquired an electron microscope, which opened for Fawcett new ways to explore with intimacy the fine structure of every tissue and organ.[3] These separate but linked events paved his way towards modernizing the teaching of microscopic anatomy and allowed him to establish an active pre- and post-doctoral program guiding the research of many trainees who came to Harvard to learn about electron microscopy. Twenty-nine of his former students and post-docs have become medical school professors and 30 others have become chairs of departments of anatomy and/or cell biology in the U.S. and abroad. One of those students was Mario H. Burgos, who learned electron microscopy with Don Fawcett and brought the first electron microscope, donated by the Rockefeller Foundation, to the Medical School of the University National of Cuyo in Mendoza, Argentina. Burgos trained a dozen researchers in the field of cell and tissue ultrastructure and I was one of those who benefited from this intellectual food chain. This kind of networking pattern can be seen in other types of situations. In 1995 I was invited to give a talk at the Congress of Mastology in Natal, Brazil, which is where I met Dr. Ismael D.C. Guerreiro da Silva. He had obtained his PhD based on his work in angiogenesis, and after talking with him about his work and aspirations I invited him to take a postdoctoral fellowship with us in Philadelphia. We published several papers[4-9] in the two years he worked in our laboratory. When he went back to his country he translated his newly gained experience in cellular and molecular biology to his laboratory, turning it into a source of inspiration for dozens of young investigators and postdocs who were the new generation of scientists in Brazil. I have had the pleasure of mentoring two of Dr. Guerreiro da Silva's students, one of whom recently received her PhD. The intellectual food chain continues.

Vocation

The scientific life is unlike any other. Scientific research is not simply a "job," rather a calling to discover the beauty of truth. Besides the elements of curiosity and an urge for understanding which are the characteristics of a scientist, one element that on many occasions is not very well considered or described is the idea of vocation. Vocation, meaning "a summons or strong inclination to a particular state or course of action,"* stems from the Latin word *vocare*, which means "to be called." While most of the time this word is used in its religious sense, "a divine call to the religious life," as in answering the call towards priesthood, I think this concept also applies to the scientific apprentice.

An individual may have the adequate intellectual acumen, or IQ, and be full of curiosity about nature's barest mechanics, but not truly fit to embark on the journey of scientific apprenticeship. To be a scientist requires commitment and a disciplined mind because good science that will in the end be useful for other human beings requires time, skill and patience. Therefore, I feel that personal motivation, or a sense of vocation, should be an important factor in considering a life in the scientific field.

I want to emphasize that the commitment required to pursue a career in science, while lofty, should not be easily dismissed. Many years of training, understanding the basic sciences and the knowledge that needs to be

---

*According to Merriam-Webster Online Dictionary 2009. Accessed 20 December 2009. http://www.meriam-webster.com/dictionary/vocation

communicated and utilized are important to science as a whole, and this needs to be imprinted in the minds of those who feel they are called to be scientific researchers. The age at which one decides to become a scientific apprentice is irrelevant, and for that reason age should not be a limitation in the scientific endeavor. It is not unusual to see researchers who are still going strong in their 70s and 80s being awarded grants, something that would have been unheard of three decades earlier. As it has been well explained by Jocelyn Kaiser,[10] because the biomedical enterprise was young and most universities had mandatory faculty retirement until 1994, there were few NIH-funded principal investigators older than 70 in 1980. But by 2007, there were at least 400 of them, according to NIH data. Indeed, NIH projections indicate that grantees over 68 could outnumber scientists under 38 by 2020. The passion for doing research doesn't correlate with youth, and it is indicated by Kaiser that people who are older and continue to work in science have as much tenacity as they have passion. "The passion for research can be extinguished in people in their 40s or 50s as much as people in their 70s," says molecular biologist E. Peter Geiduschek, 80, of the University of California (UC), San Diego. He says he will "keep doing research until somebody stops him from doing it," and that he "can't imagine doing anything else."[10] This is what I call vocation and it is so important that those whose role is in the administration of science should also keep this concept in mind. I vividly recall that one of my colleagues presented a project on a research idea that would involve a partnership between his institution and a neighboring institution, but the person in charge told him that he would be too old by the time the grant renewal came around, therefore it would be better that a younger person take charge. He was probably right, but I strongly feel that dismissing older, more experienced scientists is not only a waste of human resources, but detrimental to the maturing of the field overall.

Ratio of Individuality to Teamwork and Solitude

There is a false conception that great discoveries require serendipitous inspiration. This is not to say that serendipity is absent from the experimental process, as Alexander Fleming knew all too well when he discovered penicillin. A small mistake like a fungi contamination with *Penicillium notatum* which produces an active product called "penicillin" resulted in one of the biggest discoveries of all time that has saved millions of lives. But as I have mentioned before, in order for a "happy accident" to occur, one needs to have a mind prepared to understand such unexpected observations.

Preparing one's mind for scientific observation leading to significant discoveries requires three basic things: individuality, teamwork and the ability to work in solitude. Let's examine these concepts further. What makes a research process unique is the idea generated by one individual, meaning one needs to have the drive, constancy, devotion, and persistence to originate a research idea and the vision and foresight to develop it. This requires a significant amount of effort and the stamina to persevere hours, days, weeks, and possibly years in order to pursue this singular goal. It also requires teamwork. One of the things that distinguishes modern science from the way that was practiced a century ago is collaborative work. This is mainly because science continues to become more complex, requiring a multidisciplinary approach in which people of different talents, backgrounds and abilities come together to work on a specific problem. A critical ingredient for successful teamwork is recognizing what each scientist's role is within the team, and maintaining an atmosphere of individual thought. Preserving the uniqueness of individual thought in order to apply it to the common goal is what ultimately makes a team successful.

Solitude is the other vital ingredient that makes intellectual individuality and team approaches effective, and in the end, a driving force of scientific discovery. Solitude is an environment (both physical and mental) that allows one to contemplate a problem over and over again, to absorb and generate ideas about the end result of a lengthy process. The inability to work in solitude could be a significant detriment to the scientific endeavor. The concept of solitude is what scholars have called "meditation" and was a main source of inspiration in the Middle Ages. Solitude is related to one's ability to visualize problems and their solutions, not in a singular flash of insight, but on a repeated and continuing basis. I would like to emphasize, however, the importance of combining individuality and solitude with teamwork because it is in the latter when the management of one idea from different perspectives can help illuminate the final goal of the research process.

There have been many research projects generated in our laboratory that illustrate the need for individuality, teamwork and ability to work in solitude. Without this interconnectedness these projects would not have been possible. It was the research idea (see Chapter 2) that mammary gland differentiation during pregnancy is the mechanism that explains the protective effect of pregnancy that ultimately led us to develop the hypothesis that during this process a genomic signature of the mammary gland develops, and if it is present, would be detected in the postmenopausal breast of the parous breast, but *not* the nulliparous one. This hypothesis was demonstrated in experimental animals, but to show this in humans required a different kind of thinking. We brought an epidemiologist, a statistician, a protocol coordinator, three surgical oncologists, in addition to several molecular biologists and a pathologist, to our group. This project required the skills of each of these people. The epidemiologist, for example, designed the case control study that the statistician needed in order to understand the aims of the project in terms of determining statistical power. Once these two pieces were put together, the next phase of the study required integration of a registered nurse, through the protocol coordinator, who would conduct preliminary interviews with patients to determine eligibility. The protocol coordinator also established a connection with the surgeon who would extract the tissue. After the tissue was removed from the patient the pathologist evaluated it to confirm that it fulfilled the criteria of the protocol. But in order for all of these different pieces to work together, the coordinator had to be sure that the tissue was properly collected and preserved in the right conditions before being delivered to the lab, where the search for expressed genes would take place.

This project required that the scientific apprentice learn to be a team player in a situation where no one is in the leading role, yet each is important in their own right. Any simple mistake on the part of any one player could jeopardize the entire project. Each individual member of the team needs to understand that his or her work is part of a whole that needs to be carried out properly in order to succeed. Once all the samples are collected and processed, the solitary work of analysis and elaboration of the data begins. The team will still keep working, but there comes a phase when the scientific apprentice needs to visualize the *gestalt* of the data and provide an interpretation that will become a part of the team's discourse. But first the scientific apprentice involved in this process needs to face the solitude before being able to reach a final conclusion and understand the meaning of the data.

Scholarship

One of the aspects associated with the research endeavor is the concept of scholarship. For many years, the idea that an individual should be specialized in one specific point of knowledge has been significantly emphasized. This, no doubt, has produced tremendous advances in many areas of research, which can be easily evaluated by the number of books, journals, as well as the number of papers published in the most diverse types of disciplines. This said, I would like to draw attention to the importance of scholarship. Scholarship is derived from *scholar*, which is in one sense a learned person with general vision of the whole universe. Of course, in the Middle Ages there was a limited amount of knowledge that a dedicated individual could acquire, as well as a limited number of books with very well defined sources. Now, we have access to millions of references in different disciplines and it would be impossible for one individual to have a complete understanding of all of them. Even though I think it's crucial to have a deep knowledge of one particular discipline, I believe it is equally important to have a grasp on a wide variety of subjects. There are a lot of reasons for this, one of which might be the simple joy of intellectual richness. But the more pressing reason to cast one's scholarly net wide is because it is our responsibility toward humanity to do so. This concept is probably the most important component of the 21st century science. As opposed to the 19th and 20th centuries, where scientists pursued one idea which they maybe followed for the rest of their lives, now scientists are urged to translate that knowledge and make it useful to society, or to medicine if you are in the biomedical sciences. There are certain diseases like cancer and diabetes that we have gained a greater understanding of, yet many facets remain obscure. Therefore the need for translational thinking is part of the scholarship I'm talking about.

As part of this scholarship concept the scientific apprentice should be interacting with many different disciplines. Now we see that the most important papers published in the higher impact journals require more than one discipline, fully embracing the concept of multidisciplinary research. Multidisciplinary approaches are prevalent in medical schools, as well as in the way we tackle certain problems of morphology and function, to how genes are activated and relate to the patient, all the way from prevention to cure. This kind of conceptual elasticity requires scientists with broader training. The ability to master not only one discipline but to also understand the importance of other disciplines, and how to find a bridge that will connect their own discipline to other areas of science, is the main challenge of the budding scientist of this century. When two or more disciplines are paired together they must also be able to integrate into society in general, at which point a more complex picture emerges. Now we have a great need to overcome the deficiencies of any particular field of science in order to apply their findings to the whole. Meaning that if we discover something related to one disease, cancer for example, we must realize that we are dealing not only with cancer but also with many other diseases that may relate to sociology, economics or the environment. Not only is it important to understand disease and how to treat it, but we also need to understand the sociological, psychological and economical consequences that disease brings to society and the individual in particular. In order to finalize this concept it is extremely important that scientists be formed in a scholarly way.

In the past, scientists were only expected to know about science, but I think it is imperative to also have exposure to art and the humanities. All of these elements should be part of the training process of every individual, not just the scientific apprentice. It is expected that a college education should provide this kind of humanities background, but all too often it is

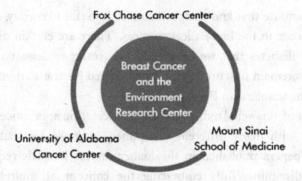

Fig. 1. Structure of the FCCC Breast Cancer and the Environment Research Center. The NCI-NIEHS supported Fox Chase Breast Cancer and the Environment Research Center is one of four consortiums working together to investigate the relationship between the environment and breast cancer. The Fox Chase BCERC is a consortium between FCCC, University of Alabama at Birmingham and Mt. Sinai School of Medicine.

not emphasized enough. Competition to get into certain medical schools is so great that students are urged to be highly scientifically and technically prepared (and rightly so), yet the effect is that other areas of scholarship, especially those related to the understanding of society as a whole, are pushed to the side. An example of this deficiency could be seen when the National Institute of Environmental Health Sciences (NIEHS), in collaboration with the National Cancer Institute (NCI), issued a Request for Applications ((RFA) #ES-03-001) in October 2003 for developing the Breast Cancer and the Environment Research Centers (BCERCs). We successfully competed by integrating three different institutions: the Mount Sinai School of Medicine in New York, the University of Alabama at Birmingham and the Fox Chase Cancer Center in Philadelphia, PA (Fig. 1). Through this program we have integrated scientific information on histologic, pathologic, cellular, and subcellular changes that occur in normal mammary gland tissue across the life span and compare this with exposure-induced changes. In addition, we conducted a focused and coordinated epidemiologic study in cooperation with two other centers organized in the same fashion for determinants of puberty in girls, with attention paid to the timing of breast development. Finally, we integrated all these data on the development and life span of the mammary gland to design public health messages aimed at educating young girls and women who are at high risk of breast cancer. As such, this type of program posed a significant challenge when it came to the meshing of disciplines in order to provide a clear message to the community.

Concluding Remarks

Part of the process of scientific research is the journey from scientific apprenticeship to researcher. As I have indicated, on some occasions it is very difficult to determine who will ultimately be at home in this field, but curiosity and a desire to understand are early signs. Motivation, dedication and the feeling that this is what one really wants to do are very important questions that need to be considered early on in the formation of the scientific apprentice. It is worth considering that while individuality in the scientific process is relevant, this individuality needs to be nurtured with teamwork, understanding and respect for others' ideas, and the ability to interact with colleagues. In addition, the ability to work in solitude in order to have a better understanding of one's ideas and mental processes is also important. Although the path of the scientific apprentice is a lonely and arduous one, it is of value to recognize those mentors that gave us the small tools that made our brains sensitive to the spark. The spark is the moment that turns the research idea, or the theme, into the engine that drives the life of the scientific apprentice. Finally, a sense of scholarship is needed so one can be prepared to transcend their own discipline and go further as a unique and irreplaceable human being, in harmony with their environment.

References

1. Russo IH, Tewari M and Russo J. Morphology and development of rat mammary gland. In *Integument and Mammary Gland of Laboratory Animals*, Jones TC, Mohr U and Hunt RD (Eds.). Springer-Verlag: Berlin, pp. 233–252, 1989.
2. Cajal SR. Mi Infancia y Juventud. Espasa Calpe SA, Spain.

3. ASCB Profile. Don Fawcett. *The ASCB Newsletter*, **23**, 2000.

4. Silva IDCG, Salicioni AM, Higgy AM and Russo J. Tamoxifen, and antiestrogenic substance, induces downregulation of CD36 gene. *Cancer Res* **57**: 378–381, 1997.

5. Russo J, Hu YF, Yang X, Huang Y, Silva I, Bove B, Higgy N and Russo IH. Breast cancer multistage progression. *Frontiers in Bioscience* **3**:944–960, 1998.

6. Srivastava P, Silva IDCG, Russo J, Mgbonyebi OP and Russo IH. Identification of genes differentially expressed in breast carcinoma cells treated with chorionic gonadotropin. *Int J Oncol* **13**:465–469, 1998.

7. Silva IDCG, Hu YF, Russo IH, Ao X, Salicioni AM, Yang X and Russo J. S100P Ca^{+2}-binding protein overexpression is associated with immortalization and neoplastic transformation of human breast epithelial cells *in vitro* and tumor progression *in vivo*. *Int J Oncol* **16**:231–240, 2000.

8. Russo J, Hu Y-F, Silva IDCG and Russo IH. Cancer risk related to mammary gland structure and development. *Microscopy Research and Technique* **52**:204–223, 2001.

9. Torres Schor AP, Marinho de Carvalho F, Kemp K, Silva I and Russo J. S100P calcium-binding protein expression is associated with high risk proliferative lesions of the breast. *Oncology Reports* **15**:3–6, 2006.

10. Kaiser J. The graying of NIH research. *Science* **322**, November 2008.

Suggested Readings

Abbott A. Hidden treasures: Turin's anatomy museum. *Nature* **455**:736, 2008.

ASCB Profile. Don Fawcett. *The ASCB Newsletter* **23**:5–8, 2000.

ASCB Profile. George Palade. *The ASCB Newsletter* **13**:8, 2000.

Baker M. Being patient. *Nature* **455**:586–588, 2008.

Bhattacharjee Y. NSF, NIH emphasize the importance of mentoring. *Science* **13**:1016, 2007.

Blobel G. George Emil Palade. *Nature* **456**:52, 2008.

Bohannon J. Vatican science conference offers an ambiguous message. *Science* **322**:1038, 2008.

Bova B. Eternal verities, eternal questions. *Nature* **404**:439, 2000.

Breve historia de la genetica, de Darwin a chargaff. *Medico Interamericano* **19**(8):354–355, 2000.

Cahlupa LM. Balancing research and teaching. *Science* **285**:2073–2078, 1999.

Carmody J. Celebrating science. *Nature* 412:383, 2001.

Cohen J and Enserink M. HIV, HPV researchers honored, but one scientist is left out. *Nature* 322:174–176, 2008.

Collins FS and McKusick VA. *Science* 321:925, 2008.

Deary I. Why do intelligent people live longer? *Nature* 456:175–176, 2008.

El futuro empezo ayer. *Medico Interamericano* 19(8):355–356, 2000.

Geherels T. Brains, courage and integrity. *Nature* 404:335, 2000.

Gibbons A. The birth of childhood. *Science* 322:1040–1043, 2008.

Gieryn TF. Who scientists are now. *Science* 322:1189–1190, 2008.

Hughes R. Visionary homebody. *Time*, February 22, 1999.

In brief: Where they stand on science policy. *Science* 322:518–522, 2008.

Jaffe S. The state of scientists' salaries. *The Scientist*, pp. 21–25, September 22, 2003.

Kaiser J. The graying of NIH research. *Science* 322:848–849, 2008.

Kirkwood TBL. Healthy old age. *Nature* 455:739–740, 2008.

Lemonick MD. Was Einstein's brain built for brilliance? *Time*, p. 54, June 28, 1999.

Miller G. Students learn how, not what, to think about difficult issues. *Science* 322:185–187, 2008.

NIH soon to be leaderless. *Nature* 455:569–572, 2008.

Norenzayan A and Shariff AF. The origin and evolution of religious prosociality. *Science* 322:58–62, 2008.

Normile D. Congress passes massive measure to support research education. *Science* 317:736–737, 2007.

Passionate Science. The 150th Essay Committee, *Science* 282:1821, 1998.

Russo E. Scientific mechanisms, past and present. *The Scientist*, p. 14, December 6, 1999.

Schekman RW. George E. Palda. *Science* 322:695–696, 2008.

Schenck DJ. On the seven paths to knowledge. *American Laboratory*, August 2007.

Secord JA. Quick and magical shaper of science. *Science* 297:1648, 2002.

Seitz F. Decline of the generalist. *Nature* 403:483, 2000.

Silver LM. What are clones? *Nature* 412:21, 2001.

Szathmary E and Szamado S. Language: A social history of words. *Nature* 456:40–44, 2008.

Vyakarnum S. The innovative brain. *Nature* 456:168–174, 2008.

Waldram J. Brian Pippard. *Nature* 455:1191–1192, 2008.

Watson JD. Honest Jim: The sequel. *Nature* 413:775–778, 2001.

Zeki S. Abstraction and idealism. *Nature* 404:547, 2000.

Chapter 2

Research Ideas

Introduction

Finding one's area of interest, or research subject, is one of the most arduous tasks a young scientist will face, yet one of the most important in that it will define the nature of a scientist's work for the rest of his or her career. For some scientists, it is a singular idea, hypothesis or discovery that defines their entire life's work, while others will make multiple contributions to their field. Mendel's name will forever be linked to the discovery of the hereditary laws, as the names Watson and Crick are synonymous with the double helix. Robert Koch and Louis Pasteur on the other hand made many landmark contributions over the course of their careers, many of which opened entirely new chapters in medicine. It is important that a scientific apprentice not be overwhelmed by the achievements of their predecessors but keep in mind that even the most revered giants of science at one time struggled with the same uncertainties in finding their intellectual niche. In that sense, the scientists of the 17th century weren't so different than those making their way in the 21st century.

The Birth of a Research Idea

While occasionally inspiration for an idea falls serendipitously into a researcher's mind, often ideas begin with suggestions made by a mentor, or the principal investigator of a laboratory. Over time, as an apprentice continues working in a lab, he or she might make an association based on the surrounding research milieu and from there go on to awaken a deeper interest in a given topic.

Sometimes it is life experience that draws students to a particular area. Many scientific apprentices have indicated to me that their desire to study breast cancer stemmed from the loss of a mother, an aunt, or a grandmother to the disease. The emotional and fundamentally instinctual desire to alleviate suffering, as much as the introduction of a challenging idea, could mark the beginning of a lifelong path.

Whatever the origin, the birth of a research idea is, in the end, the result of triple processes working together: reading, writing and experimentation. Only when carried out in tandem, the research endeavor, and the discovery of oneself as a scientist, becomes possible.

Reading

As in even the most basic learning, reading is the scientific apprentice's first tool in learning the trade. There are several types of reading: general, review, original and technological.

General papers, or publications, tend to describe the big picture of a given biological process or concept. Not only do they serve as a means of introduction into a particular area, but they may also provide the human perspective on scientific issues. For example, an article in a popular magazine might focus on the economic consequences of disease and therefore broaden the reader's conception of who is affected by it. If a scientific apprentice is inclined towards clinical research, an article describing the manifestation of breast cancer, from onset through to the patient's final days with metastasis, and the varied presentations based on age and cancer type, could be an eye opener to the complexity of this disease. A budding sociologist might look at the same article and be captured by disparities in treatment and prevention. These types of articles also provide insight into the multidisciplinary nature of science, which is so difficult to grasp early on. For some, it is not an easy task to see clearly how statistics, sociology, economy and biology work together to form a particular problem (or solution). Perhaps this is why specialization has become rampant, especially in science. I don't in any way mean to diminish the importance of highly developed expertise, but often in the race for achievement, the crucial frame of contextual understanding and knowledge necessary for the growth of a young scientist is weak.

I have written many general interest articles over the years and what I have enjoyed most about them is the feedback I've received from readers. While these pieces might appeal to a wide audience from various backgrounds, they are not easy to write. General articles are demanding, linguistically because they should be lively and engaging, but conceptually as well. P. Michael Conn, in his article "Make science Relevant, Human, and Clear" writes, "the first rule in communicating science to the public is: Make it relevant. Communicate in human terms. Explain how issues are important to people. The second is a correlate: Avoid confusion. Don't jargonize. That's just good writing. The final is: Do it. Take the time. Make the commitment." Conn continues,

… the U. S. public first noticed science after World War II, when the impact of the atom bomb brought scientists to the forefront… The promise had become reality: penicillin, polio vaccine, and space exploration. The magic of being able to prevent and cure disease was upon us; everyone's life was directly affected. Our major concern was that with science and technology curing our diseases, cleaning our homes, and exploring our universe, we would have little to do. As we communicate the success, awe, and wonderment of science, the wonders themselves evoke questions and worries. Consider, for example, from 1972, the Tuskegee Study of Untreated Syphilis in the Negro Male. Ethics was not a popular topic in biological journals in the 1970, but today many journals raise issues of ethical concerns; scientists are pensive and often divided themselves on issues that, at one time, were not even discussed. There is increased contemplation about the problems of socially applied science. The potential problems are now matters of public and legal concern.

Conn's comments serve as a reminder that science reaches far beyond the lab. As a member of society the scientific apprentice is a member of the general article's audience and can benefit from information presented in a global context.

Review articles serve a different purpose entirely. Their purpose is to bring fresh meaning to a specific scientific topic by exploring possible interactions and translational ideas. A good review article will also address future directions for the research being discussed. In addition, it should provide an overall framework of references to what has been accomplished up to the point in history in which the scientific apprentice is living. If done well, the review article will go beyond simply discussing methodological approaches and address the broader relevance and future of the research. The review article is an important tool for the scientific apprentice. It is in writing them that one begins to look beyond the print and deep into the ideas themselves. Writing about the future directions of a given topic is the most challenging aspect of the review paper, but also the most rewarding because this is where one begins to grow. Creative work begins with the review paper; the writer must present novel ideas which can act as seedlings towards the development of new research. The ideas explored in the review article can also be a measure of the writer's capacity for innovative thinking, and may assist the scientific apprentice in finding their own niche within research.

Shai Vyakarnum in his article, "The Innovative Brain," indicates that innovative, or risk-taking approaches could be considered abnormal expressions of behavior in association with substance abuse and bipolar mania. But Vyakarnum presents a new perspective to these observations:

> Although the innovative quality in science cannot be compared completely with that of a gambler, it reflects a behavioral index of risk-seeking or risk tolerance. Greater rewards (as well as greater losses) are available for those who bet more. The betting behavior decreases with age across the lifespan therefore the innovation of ideas in science could be initiated a younger age when higher risk-taking is likely possible.

In other words, the scientific apprentice is poised to take those intellectual risks. In order to take those risks — to explore new ideas, correlations, scientific possibilities — a scientific apprentice must be in a safe and supportive research environment, at the beginning provided by a good mentor, and later on by an institution heavily supportive of his or her scientific research.

If general and review articles help form the basis for the development of research ideas, it is in reading original scientific publications that the scientific apprentice's mind begins to grasp the full research process. A critical skill for the scientific apprentice to hone is the ability to distinguish conceptually challenging ideas from the simply technically challenging. Gross and Harmon write in their enlightening article "What's Right About Science Writing" that the "scientific article since its inception in the 17th century, and remains today, the most effective medium for honestly communicating the cognitively complex knowledge scientists create."

Several pieces of advice could be given to the scientific apprentice regarding how to go about reading scientific papers: read only peer-reviewed papers published in high impact journals (more on this later), or read the work of long-established laboratories from highly regarded institutions. These are valid suggestions, and should be kept in mind, but should not be limiting factors in what a scientific apprentice chooses to read. While rare, there have been instances in which a paper published by a highly respected lab was recalled due to inaccuracies, even fraud. There is an abundance of literature about misconduct in science. The most crucial piece of advice I would give to a young scientist, in regards to their reading habits, is to read critically. Be active and engage with the material. Over time, the scientific apprentice will develop an instinct for originality in ideas, methods and conclusions. The scientific apprentice should not rule out the work of obscure scientists from lesser-known universities in far corners of the world. Science happens everywhere and the scientific apprentice might be surprised to find an innovative hypothesis posed by an unknown researcher in a place that due to economic or language barriers, might not be able to compete within mainstream science. Original articles can help spark an idea that may lead to a piece of knowledge that will ultimately define an apprentice's main research idea. One can only wonder how far Darwin would have gone had Mendel's publication on genetics come to light in his own day.

Technical literature has an important function in the development of the scientific apprentice as well. Reading technical articles is the best way for a young scientist to discover how the mechanical components of an experiment are carried out and what they mean for the larger experiment.

Writing

Scientists are by necessity writers. Academic scientists write exams, grant applications, and research papers; their industrial counterparts compose company reports and patent applications. As Gross and Harmon write, "the successful enterprise of science depends crucially on the efficacy and honesty of its communicative practices." Therefore, in the life of the scientific apprentice, writing and reading should exist together, as part of the continuum of scientific practice. Most of the time, however, writing is pushed to the periphery, treated as necessary grudge work to finish as soon as possible. When writing scientific papers, often the truly creative part — the discussion and interpretation of the results — does not come until later, when the experiment is finished and the results obtained. That's not to say that writing doesn't have an important role beyond the all-important scientific paper, which in many cases is the final act and endpoint.

Writing review articles, such as those discussed earlier, is one of the best ways to develop this essential skill. Writing these kinds of analytical pieces forces the mind to think actively and critically about specific studies and how they relate not just the current wave of scientific thought, but to the scientific apprentice's own interests as well. Review articles can also train the young scientist to see in what direction any given experiment needs to go in order to truly have an impact. This type of exploration into the future of certain studies is sorely lacking in the plethora of review articles published everyday. In searching the published literature the scientific apprentice can begin to understand the commonalities and differences between different researchers working in the same field. This in itself is an essential component of the writing process and discovering how knowledge on a particular subject emerges is just as rewarding as comparing different techniques and experimental models. Equally important, the scientific apprentice is learning to see beyond what is written and discover the loose ends that extend beyond what is known. Engaging in the reading and writing process early on prepares emerging scientists to elaborate their own subject of study and in the most crucial, demanding and even enriching task of a scientist's career, grant writing, a topic I'll discuss in following chapters.

Another common, yet often over-looked, component of any paper is visual representation. Be it a graph, table, photograph or illustration, at times it is necessary to move beyond words and express more subtle or complex ideas with an image. For Cajal, the drawings he made of neurons were instrumental in helping society understand his observations. In an age before photographs and powerful microscopes, the best and only way to show the world what you saw in your mind was to draw it. Needless to say, the art of visualization has grown exponentially since Cajal's time, especially in the last ten years. Once the scientific apprentice becomes proficient in some of the many digital imaging tools available today, he or she will have more dexterity in communicating their findings.

Experiments and Benchwork

Scientists need to learn not only a lot of technical skills early on, but also the reasoning and principles behind them. Working at the bench can be tedious and requires a special kind of tenacity. Those who enjoy lab work the most from early on are most likely to be well suited for a life in science, and the sooner an individual is exposed to experimentation, the better. Mastering the tools of research and understanding how they work and why will also be of help when trying to determine an area of research interest. For example, learning the morphological sciences and the techniques used in studying developmental biology is excellent preparation for understanding the pathology of diseases and exploring their molecular and biochemical pathways. Over the years, as I have watched many apprentices learn the skills of research, I have observed that those who begin with the morphological sciences and become familiar with identifying tissue under the microscope and then move onto the more complex molecular end have an easier time grasping the totality of the related concepts.

A scientific apprentice must invest as much time as possible in fully understanding the basics — beyond the textbook. In being utterly at ease with basic laboratory techniques and the principles that drive them the apprentice will then not only be able to quickly learn and adapt to new techniques and information, but also be bolder in their problem solving approach. Having a broad base in biomedical sciences does not preclude the eventual narrowing of focus, but familiarity with other disciplines such as epidemiology, mathematics, physics and even economy, sociology and literature will make it much easier to work alone, be part of a team, and eventually direct one.

While the approach I discuss above is ideal, the problem is that there are not many mentors who are able or willing to orient a scientific apprentice. Our current system of tutoring is narrowly focussed — from the beginning stages all the way to specialization. In reading the work that appears in high quality journals such as *Nature, Cell, Science,* and others, it is clear that high quality mentorship and training does exist. Peer-reviewed journals present work that requires a comprehensive approach to solving biological problems. Our collaborative work on the role of SATB1 reprogramming gene expression to promote breast cancer metastasis is an example of this kind of approach, in that it required a significant interaction between molecular and cell biology, informatics, experimental and human breast pathology with clinical outcome of the disease, all of which contributed to the paper's publication in *Nature*.[1]

A Formula for Daily Practice in Research

Like most things in life that require skill and thought, science is something that must be worked at daily. There are practical things that can help move this process along more fluidly. Do things that are hard for you at the time of day in which you are at your mental peak. If you are a morning person and writing is your least favorite activity, write first thing in the morning. Likewise, save those things that you enjoy or come most naturally to you for when you find yourself losing focus. It is easier to do things that require less thought or motivation when you are sluggish.

You must never entirely cease benchwork just because you are writing a paper or "thinking." I cannot stress this enough. Benchwork is part of the continuum of scientific practice and an important part of maintaining an active mind. Because I am at my best in the mornings, that is when I write most of my grants and papers. But even at this advanced stage of my career, where I have barely enough time to work at the bench, I still force myself to use the microscope every day as a way to connect to the practical side of the lab. I no longer have time to culture cells or extract RNA, but for me (and because I am at my core a pathologist) the microscope serves as an extension into that environment. It is important for an apprentice to get into the habit of alternating the often grueling physical work with the more cerebral written analyses. Even the biostatistician whose work exists entirely in a conceptual and mental realm might benefit from an occasional foray into the lab. I have had a bioinformatician tell me that he would love the opportunity to spend a day at the bench so he could better understand where all the data actually comes from.

While benchwork is getting more sophisticated and expensive, reading and writing in this electronic age have never been easier. Almost every article published is available through the web, and added to this cornucopia of information are a growing number of "data mining" databases that are accessible to all and out of which theoretical or "cyber manuscripts" can be generated. This has allowed complex processes to be studied more closely because information, such as gene pathways, are now stored on online databases where anyone can take a look and test their hypothesis. This is a critical area that has advantages and disadvantages. One advantage is that due to the availability of gene databases, it is now easy to compare them with what your laboratory has generated and get an idea as to whether the data are similar or discordant. One might even get better, or altogether new, conclusions about similar material with equal methodological approaches. These online databases might even help establish contact with scientists in other fields. This kind of collaboration is what science is all about. The disadvantage of making these databases so accessible, however, is that anybody can use them to generate new conclusions even if they have never generated original data themselves or set foot in a wet lab. Whether this is right or not isn't for me to say, but I feel uneasy when others get conclusions from data generated by others without honestly and openly communicating with the researcher who originated them in the first place.

Within biological sciences, physical experimentation is still necessary. All the above-mentioned data originates from the accumulated efforts of individuals working alone as well as part of a team. We are not anywhere near the stage where we can solve biological problems — cure disease, find cause and preventive measures — using cyber technology alone. It is up to the scientific apprentice to learn how to connect the dots; to use the methodology of laboratory work (RNA extraction, microarrays, Western blots, cell culture) to generate the complex data that may some day find answers to some of science's most difficult questions.

It is well known that science thrives in large part thanks to government funding. In periods of economic restraint the need to justify the applicability of one's research is intensified. Competition gets tougher as many researchers from all over the world are competing for access to the same limited pool of money. While those who go into research do so for the love of science, knowledge, and human advancement, there is a harsh reality to face: research is expensive. The scientific apprentice must learn to discern for him or herself what is pure science — the generation of knowledge for the sake of knowledge itself — and science which while enlightening, also serves an applicable function. Was the time Koch spent perfecting the right stain for the *Myobacterium tuberculosis* wasted? At first it might have seemed so, for it wasn't apparent what good it would do. But without that stain, the bacillus would never have been discovered as the etiology of a rampant and devastating disease. One of the fundamental problems of scientific study is that it is impossible to predict the outcome. Was Golgi aware that his method of argent impregnation would be used by his rival Cajal who would go on to make one of the most significant discoveries in the history of neurology? Golgi never imagined that his technique, which consisted of saturating fragments of brain tissue for several days with a solution of potassium bichromate, and later treating the fragments with silver nitrate so that a deposit of silver bichromate could be picked up by certain nerve cells to the absolute exclusion of the others, would pave the way for someone else's discovery. Golgi didn't see what Cajal saw. The scientific apprentice must therefore focus their attention on the research idea and not cloud their thinking with peripheral concerns such as: *Will anyone like it? Is the idea accepted enough by the establishment? Will it win the Nobel prize?* If there is a genuine desire to research a particular topic, the rest will come on its own.

Training Grounds for Developing and Testing Research Ideas

One of the questions that emerged from the previous section is how to evaluate for oneself whether a research idea is headed in the right direction, is innovative and has the potential to be successful. Use seminars as an opportunity to present your data and pay attention to your peer's and mentor's feedback. This exchange can be helpful when writing your discussions or developing new ideas. Never postpone a seminar because you believe that you do not have enough data. Present your data, your ideas, challenge the audience of your peers, and listen for comments that can trigger new ideas, or an interpretation that may have initially escaped your notice. Never be afraid of critiques, only of silence. If your peers say nothing it could indicate indifference toward you or your work, or that you are not in the right environment. Science is carried out by human beings, so it is naturally constrained and potentiated by the scientific apprentice's cultural and historical milieu. For example, it is not possible to determine whether Mendel's research idea began as clearly as we imagine, or whether it started, as indicated by Henig, as a basic concept to solve a practical problem: finding out how hybrids reproduce and determining what will happen to their progeny. The idea was to determine why hybrids produce plants similar to themselves, while on other occasions they look more like those of a previous generation, not to establish the law of inheritance.

In pursuing a research idea we must always remember that the main sources of knowledge are observation, experimentation and inductive and deductive reasoning. Claude Bernard further adds that the researcher cannot be oblivious to natural phenomena, or believe in determinism. Our mission is to ascertain the mechanisms by which they happen and not the reasons for their occurrence.

There are classic books that must not be ignored like the *Novum Organum* by Francis Bacon that trigger the thinking process, but do not teach the scientific apprentice how to have a sense of discovery. Descartes' *The Methods*, however, offers practical approaches, such as solving a problem by dividing it into parts and starting from the most simple and working towards the more complex. The best example of this advice is seen in Cajal's work. He started by studying the nerve cell interactions in single organisms which ultimately led him to understand complex structures such as the eye of the rabbit. This kind of advice by way of example has been extremely useful in our research endeavor.

The Evolution of a Research Idea

A research idea may originate from an isolated event, as we saw with Fleming's discovery of penicillin, or from a long process of trial and errors. It is interesting to note that Fleming made the discovery of the lysozyme in 1922 when a drop of mucus leaked from his nose while he was working with some bacteria. This drop caused the bacteria to disappear a few hours later. He had unwittingly discovered a natural substance found in tears and nasal mucus that helps the body fight germs. Although others pursued study of the lysozyme independently, he made the important initial discovery that a substance could kill bacteria without negatively affecting the human body. Another accident took place six years later when Fleming was sorting through a number of glass plates that had previously been coated with *Staphyloccus* bacteria and noticed one of the plates had developed some mould. The mould was in the shape of a ring and the area around the ring seemed to be free of the bacteria *Staphyloccus*. The mould was *Penicillium notatum*. Further research on the mould found that it could kill other bacteria and that it could be given to small animals without any side effects. This discovery was rescued ten years later at Oxford by Howard Florey, Ernst Chain, and a junior member of the team, Norman Heatley, who isolated the bacteria-killing substance found in the mould, what we now know to be penicillin. Florey got an American drug company to mass-produce it and by D–Day (June 6th 1944), enough was available to treat all the bacterial infections that had broken out among the troops. In 1945, Fleming, Chain and Florey were awarded the Nobel Prize for medicine but unfortunately Norman Heatley, who had played a significant role in the purification and development of the substance, was not included in this honor. This case calls to mind William Osler's words: "In science the credit goes to the man who convinces the world, not to the man to whom the idea first occurs." While Heatley did not discover the monumental drug, it was his tenacity and originality that helped make it as ubiquitous as it is today.

Not all discoveries are serendipitous, in fact most are based on thorough analysis. In the early 1960s, a researcher named Jerome Horwitz at the Michigan Cancer Foundation in Detroit, Michigan, obtained a grant from the National Cancer Institute to design a group of compounds called dideoxythymidines that were designed to look like nucleosides, the building blocks of DNA. In theory, these nucleoside analogs would substitute themselves for real nucleosides, thereby interfering with formation of DNA molecules. The idea was that without more DNA, the cancer cells would simply stop duplicating. The compound synthesized by Horwitz, AZT, turned out to be completely ineffective against cancer cells, and so toxic that Horwitz did not even bother take out a patent. In this case there was an excellent idea, an investigator who got a grant based on a novel concept and a product that in light of the research was basically useless. And for almost 20 years it remained so. It wasn't until the 1980s came along and with it the AIDS virus that this compound was able to make an adequate inhibitor of the viral replication mechanisms. Dr. Horwitz received recognition for this, but the Michigan Cancer Foundation did not have a patent on the drug and on March 20, 1987, the AZT was licensed as an AIDS drug as "Retrovir."

On other occasions the origin of a research idea can be traced back many years. Figure 1 depicts the chronology of a research idea that is still in process. In 1972, when Irma H. Russo and I began working on the prevention of human breast cancer, we found many problems in the literature regarding how to collect data from human material in order to shine new light on the cancer process. It became clear to us that if we wanted to prevent breast cancer we needed to have a good grasp on how cancer was formed, meaning understanding the pathogenesis of the disease. We decided to start working on the pathogenesis of breast cancer, but again we faced the problem of not having the right methodology to see beyond what was already reported in the literature. It was at this point that we decided to start using a single experimental model by studying mammary carcinogenesis in rodents. Using the basic model developed by Charles Huggins in 1960 consisting of intragastric administration of DMBA (7,12-dimethylbenzanthracene) to young

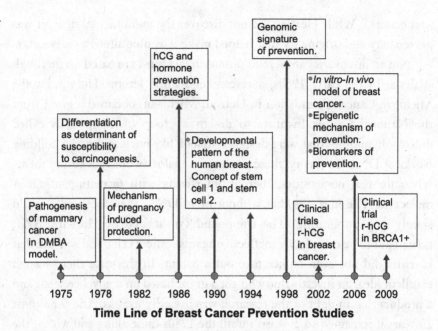

Time Line of Breast Cancer Prevention Studies

Fig. 1. Time line of the breast cancer prevention studies indicating the main landmarks of this research idea that started in 1975 and is still developing.

virgin Sprague Dawley rats, mammary carcinomas were formed in almost 100% of the animals. These studies have led us to the conclusion that the carcinogenic potential of a given chemical is in great part modulated by the biological conditions of the target organ which determines its susceptibility to neoplastic transformation. In the rat this requires that a carcinogen act on a specific compartment of the mammary gland, the terminal end bud (TEB), a club-shaped undifferentiated structure found at the peripheral margins of the developing mammary parenchyma in young virgin rats (Figs. 2a–c and Figs. 3a and b). When we saw that the terminal end buds of the immature mammary gland were the sites in which the cancer started, this lead us to the discovery of the intraductal proliferation that was featured on the cover of the *Environmental Health Perspective Journal* published by the National Institute of Environmental Health Sciences in 1996. It took several years of patiently studying the rat and the mouse model to evaluate the importance of the branching pattern and lobular development. The ease of the rat and

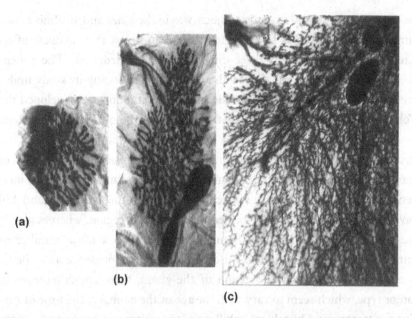

Fig. 2. Whole mount preparation of the abdominal mammary gland at (a) 21 days of age, (b) 35 days of age, (c) 55 days of age. Toluidine blue ×10. (Reprinted from: Russo, J. and Russo, I.H. "Biological and Molecular Basis of Breast Cancer", Springer-Verlag, Heidelberg, Germany, 2004.)

Fig. 3. Whole mount of abdominal rat mammary gland: (a) control, (b) after 35 days of DMBA administration. Toluidine blue ×4. (Reprinted from: Russo, J. and Russo, I.H. "Biological and Molecular Basis of Breast Cancer", Springer-Verlag, Heidelberg, Germany, 2004.)

37

mouse mammary gland as study subjects was in their size and an almost two-dimensional structure as opposed to the human breast that, because of its size, would have required three-dimensional reconstruction. The rodent gland fits nicely on a glass slide with a cover slip facilitating its study under the stereomicroscope or the light microscope. In addition, we developed the Toluidine blue stain to differentiate the epithelia and the stroma in the whole mount, which gave us wonderful differentiation over other cells in the stroma, such as the mast cells. Although TEBs are present in the six pairs of mammary glands, tumor development does not occur randomly. Tumor incidence in animals treated with a carcinogen between the ages of 20 and 180 days is greater in those glands located in the thoracic region, whereas glands located in the abdominal and inguinal areas develop a lower number of tumors (Fig. 4). In addition to differences in tumor incidence as a consequence of the topographic location of the gland, there are differences in tumor type, which seem to vary with the age of the animal at the time of carcinogen treatment. Ductal and papillary adenocarcinomas are more frequent in both thoracic and abdominal glands of younger animals, whereas adenocarcinomas with a tubular pattern are found mostly in abdominal glands of older animals.

The studies on the pathogenesis of mammary cancer opened a new window to another important issue, which was the development and differentiation of the mammary gland. The development of the rat mammary gland occurs through a combined process of branching and differentiation of the parenchyma, mainly in those ducts ending in TEBs that progressively divide and differentiate into alveolar buds (ABs). These structures in turn differentiate into lobules (Fig. 5).

Although this pattern of development is common to the six pairs of mammary glands, it does not occur simultaneously in all of them, rather varies in relation to the topographic location of each specific pair. Individual structures, (i.e., TEBs, ABs, and lobules) appear similar in morphology in all the glands, however, their relative number and the general architecture of the organ vary notably from one pair of glands to another. The most notable ones are the thoracic mammary glands, since each single gland is composed of two different layers separated by connective and muscular tissue; one layer is composed of

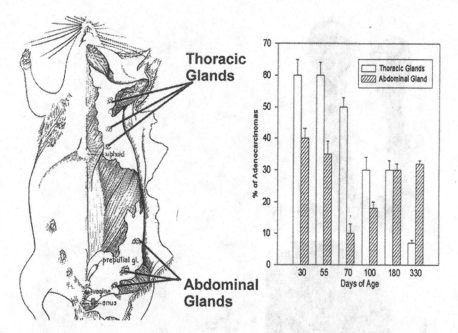

Fig. 4. Drawing of a rat showing the distribution of the six pairs of mammary glands. The fourth pair, or abdominal mammary gland, is the one most frequently utilized for morphological, cell kinetics, and tumorigenic studies because of its large size and ease of access. However, they are less susceptible to transformation by chemical carcinogens, as indicated in the histogram on the right, which shows that the higher incidence of adenocarcinomas occurs in thoracic mammary glands in those animals inoculated with the carcinogen at ages younger than 100 days. These results were obtained by inoculating virgin Sprague Dawley rats intragastrically (ig) with a single dose of 8 mg 7,12–dimethylbenz(a)anthracene (DMBA) (Sigma Chemical Co., St. Louis, MO) per 100 g body weight (bw). The carcinogen was dissolved in corn oil at a concentration of 16 mg DMBA/ml by heating in a water bath at 95°C for 30 minutes. (Reprinted from: Russo, J. and Russo, I.H. "Biological and Molecular Basis of Breast Cancer", Springer-Verlag, Heidelberg, Germany, 2004.)

more numerous ABs and small lobules, whereas the adjacent one is more extensive and contains thin long ducts ending in prominent TEBs. The abdominal glands have a markedly reduced number of TEBs, which are located exclusively in the most distal portion of the gland, whereas the middle and proximal portions show a much more differentiated appearance. The difference in number of TEBs in thoracic versus abdominal mammary glands is significant. With aging, TEBs decrease progressively and their reduction is

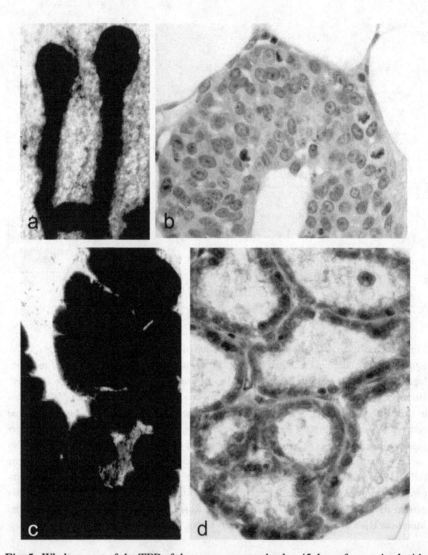

Fig. 5. Whole mount of the TEB of the rat mammary gland at 45 days of age stained with toluidine blue ×20. (b) Histological section at the tip of (a), stained with H&E, ×40; (c) Lobule type 3 of a pregnant rat mammary gland stained with toluidine blue ×20; (d) Histological section of (c) stained with H&E, ×40. (Reprinted from: Russo, J. and Russo, I.H. "Biological and Molecular Basis of Breast Cancer", Springer-Verlag, Heidelberg, Germany, 2004.)

proportional in all the glands. This reduction is mostly due to either their regression to terminal ducts, or TDs, or to a greater differentiation to ABs and lobules. The higher incidence in ductal carcinomas observed in thoracic glands is attributed to the difference in degree of development of the undifferentiated layer of this gland in comparison to the glands located in other topographic areas. This seminal work spawned the development of new concepts in the understanding of the susceptibility of the mammary gland to carcinogenesis, concepts such as the role of differentiation and rate of cell proliferation of the mammary gland at the time of exposure to a given chemical carcinogen on its binding, DNA repair and tumor incidence, and have influenced the way in which the study of mammary cancer is pursued by researchers worldwide.[2–7]

Studies have progressed from understanding the pathogenesis of mammary cancer in rodents, to the validation of this model for the study of the human disease. The study of the human breast had already been attempted by anatomists and embryologists and it was the translation of Dabelow's chapter on the human breast from the original German version that gave us the inspiration to pursue the study in humans using whole mount preparations of serial sectioning of the human breast and applying the same methodology that we did to study the rat mammary glands. Irma and I relied on the valuable assistance of Maria Kosalska and Ana Lillian Romero to prepare and obtain the tissue, but it was the systematic study performed by Irma and I that yielded the tri-dimensional reconstruction of the human nulliparous and parous breast. We received support from the NCI in the form of a three-year grant to perform these studies. Research conducted during these years has led to the demonstration that undifferentiated structures of the human breast, designated lobules type 1, are the site of origin of ductal carcinoma, the most common type of cancer,[8–10] More importantly, it was shown that the pattern of development and differentiation of the breast differs between nulliparous and parous women. Lobules type 1 are the most frequent structures present in the breasts of nulliparous women, exhibiting a high rate of cell proliferation. Their *in vitro* binding of a carcinogen to the DNA and their susceptibility to transformation in culture by chemical carcinogens are greater than those of the differentiated lobule type 3 found in the breast of parous women.[10] From those studies we were able to describe the differentiation pattern of the human breast and the difference between the nulliparous and parous women, and postulate that the protection conferred by pregnancy against breast cancer was due to the fact that during pregnancy and lactation, the breast tissue completes the process of differentiation, conferring a specific genomic profile, or signature, that makes it refractory to carcinogenesis (Figs. 6 and 7). These studies opened new and important translational possibilities by

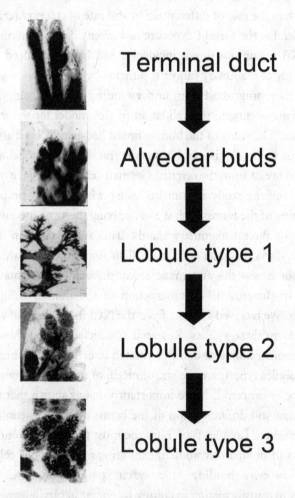

Fig. 6. Lobule types in the human breast. (Reprinted from: Russo, J. and Russo, I.H. "Biological and Molecular Basis of Breast Cancer", Springer-Verlag, Heidelberg, Germany, 2004.)

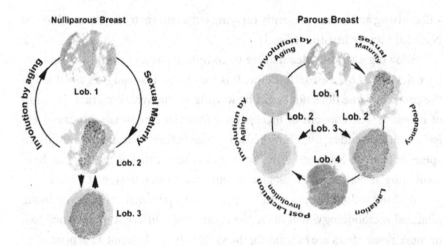

Fig. 7. Cycle of differentiation of the nulliparous and parous breast. (Reprinted from: Russo, J. and Russo, I.H. "Biological and Molecular Basis of Breast Cancer", Springer-Verlag, Heidelberg, Germany, 2004.)

using the mechanisms of cancer initiation, which led us to develop strategies for its inhibition, capitalizing on the utilization of physiological mechanisms, such as pregnancy. Based on this we decided to focus on a single placental hormone, human chorionic gonadotropin (hCG), for protection from chemically induced mammary cancer in virgin rats. The protective effect of hCG, like that of pregnancy, is mediated by the induction of full differentiation of the gland, which eliminates the undifferentiated terminal end buds that are the carcinogen's targets.[11-14] The relevance of these studies lies in the epidemiological data revealing that parous women have a four-fold lower breast cancer incidence than nulliparous women, therefore providing a physiological basis for the prevention of cancer in humans. The long-lasting collaboration and partnership between Irma and I allowed for the demonstration that the mechanism of hCG's action is mediated by a complex network of newly identified genes.[15] The discovery that hCG mimics pregnancy, that the protection conferred to the mammary gland from the initiation of the neoplastic process is equal or more efficient than that induced by the gestational process, and that it inhibits the progression of tumors, indicates that this model might be an ideal tool for breast cancer prevention and therapy.[15] This is a subject that is still evolving and we are presently carrying out a clinical trial sponsored by the National Cancer Institute (Fig. 1).

Allowing a simple idea mature to complexity is an example of how we have developed our research theme. It is easy to get lost and overwhelmed in the process if one does not begin by looking at the overall picture as a sum of its parts. The concept of increasing a subject's complexity progressively isn't such a new idea, after all we learn to add before we learn to multiply, to speak before we learn to write, so it makes sense that when approaching something as complex as biological science we do so one step at a time.

Without a doubt the scientific apprentice's personal touch, even using identical methodology, will make the same study different, even when two or more researchers are examining the same subject. It is not a surprise then that even when starting with the same question each scientific apprentice will find a different path or make a different observation. The important thing to remember is that if they are correct, the universal principle will apply throughout.

In the search for a research idea the scientific apprentice may find that the chosen theme will challenge established concepts, either because of the advent of new technology or because they approach the problem from a different perspective. To challenge an established idea or concept in the biological sciences requires first to understand the biological principle in question, but more importantly to know how the knowledge that gave way to this concept was developed. In 1970, for example, the preponderant concept was that the hyperplasic alveolar nodule in the mouse mammary gland was the precursor to cancer. This principle still holds true in the mouse and is a specific biological characteristic, just like most of the tumors originated in strains that contain the murine mammary tumor virus are pregnancy dependent. Therefore it is the hyperplastic alveolar nodules, or HAN, that are remnants of the involuted gland after pregnancy and the ones that will evolve to carcinoma. However, this type of lesion is not found in the rat mammary gland and even when the studies were done in human the results were not properly interpreted. When we started studying the DMBA induced mammary model in rats we discovered that the lesions started in the terminal end bud, or TEB, of the mammary gland were of undifferentiated structure. The observation and the quantitation of the phenomena concluded that in the rat, as opposed to the mouse, the neoplastic process starts in the ductal component, mainly in undifferentiated structures. When the TEB differentiates into alveolar buds and lobules they lose their susceptibility to neoplastic transformation, which explains why the window of susceptibility in the rat is around 45 to 55 days of age. After that age the number of TEBs decrease and become less susceptible to developing cancer. The importance of this observation is that it mimics what was observed in the human breast, where most of the cancer starts in the ductal structures. In 1978 we published a paper establishing the similarities between the rat and the human mammary glands.[16]

The lesson to learn here is that only solid observations are valid, and when a scientific apprentice is secure of their of findings they must not be afraid to present them even if it means challenging or criticizing what has been accepted up to that point. Overall, the scientific apprentice needs not only self-confidence and strength, but also needs to maintain a profound respect for those who hold opposing views. In addition, they must acquire a profound understanding of the divergent view. It is important not to confuse strength and assertiveness with arrogance. The latter does not have place in the life of the scientific apprentice.

We cannot change history when it comes to how new ideas have clashed with old concepts. Let us not forget the controversy between Cajal with Golgi, in which the former's idea of nerve cells ultimately gave way to the understanding that these cells are connected to each other through the synapses ultimately prevailed, or the discussion between Schwann's notion on the spontaneous generation of cells vs. Virchow's concept of *omni la cellule a cellule*, which was eventually accepted as a biological principle, or the polemics between Pasteur and proponents of the spontaneous generation of germs. These are but a few examples of many classic encounters among scientists. These examples may give the impression that there is a constant clash between the younger and older generation because the latter try to keep their theories intact and postulate against the younger generation's new ideas. While there is some truth to this, present data indicates that older scientist are producing more publications and generating more new ideas, as measured by grant awards, than the younger generation. This does not indicate that one generation is better than the other, but it is an indication that the scientific apprentice can learn from management of research ideas and the theme of work from those who toiled at the bench before them.

Scientific controversies, such as that between Francis Collins and Richard Dawkins on the existence of God, are healthy in that they help create a sense of what present scientists believe in or not, and how their belief influences their actions. Whereas for Collins the genetic code is the language of God, Dawkins sees it as simply part of the evolutionary process. All the parties are enriched by the discussion, and may lead some to new ideas and themes of work.

Final Remarks

In reality, the questions and themes of research are out there, it is just a matter of the scientific apprentice finding them. The only time when there ceases to be new ideas is when imagination and will are extinguished. What the scientific apprentice needs to learn is how to create knowledge, and in the same way that every nucleotide sequence encodes a message, everything that surrounds us and needs to be discovered, because discovering the secrets of the universe is part of engaging in a creative act. Any new idea, any small piece of work or knowledge created is part of the final discovery. While in the Olympic games there is only one gold medalist in each event, everyone who made it as far as the games is a winner in their own way as well. Such is the process for the scientific apprentice who needs to come to their own realization as an individual and enjoy the creative act of everyday work in the research endeavor.

References

1. Hye-Jung Han, Russo HJ, Kohwi YJ and Kohwi-Shigematsu T. SATB1 reprograms gene expression to promote breast cancer metastasis. *Nature* 452:187, 2008.
2. Russo J, Saby J, Isenberg W and Russo IH. Pathogenesis of mammary carcinoma induced in rats by 7,12-dimethylbenz(a)anthracene. *J Natl Cancer Inst* 59:435–445, 1977.

3. Russo IH and Russo J. Developmental stage of the rat mammary gland as determinant of its susceptibility to 7,12-dimethylbenz(a)anthracene. *J Natl Cancer Ins* **61**:1439–1449 ,1978.

4. Russo J and Russo IH. DNA labeling index and structure of the rat mammary gland as determinant of its susceptibility to carcinogenesis. *J Natl Cancer Inst* **61**:1451–1459, 1978.

5. Russo J, Wilgus G and Russo IH. Susceptibility of the mammary gland to carcinogenesis. I. Differentiation of the mammary gland as determinant of tumor incidence and type of lesion. *Am J Pathol* **96**(3):721–734, 1979.

6. Russo J and Russo IH. Influence of differentiation and cell kinetics on the susceptibility of the rat mammary gland to carcinogenesis. *Cancer Res* **40**:2677–2687, 1980.

7. Tay LK and Russo J. Formation and removal of 7,12-dimethylbenz(a)-anthracene-nucleic acid adducts in rat mammary epithelial cells with different susceptibility to carcinogenesis. *Carcinogenesis* **2**:1327–1333, 1981.

8. Russo J, Calaf G, Roi L and Russo IH. Influence of age and reproductive history on cell kinetics of normal human breast tissue *in vitro*. *J Natl Cancer Inst* **78**:413–418,1987.

9. Russo J, Reina D, Frederick J and Russo IH. Expression of phenotypical changes by human breast epithelial cells treated with carcinogens *in vitro*. *Cancer Res* **48**:2837–2857, 1988.

10. Russo J, Calaf G and Russo IH. A critical approach to the malignant transformation of human breast epithelial cells. *CRC Crit Rev Oncol* **4**:403–417, 1993.

11. Russo IH, Koszalka M and Russo J. Human chorionic gonadotropin and mammary cancer prevention. *J Natl Cancer Inst* **82**:1286–1289, 1990.

12. Russo IH, Koszalka M and Russo J. Effect of human chorionic gonadotropin on mammary gland differentiation and carcinogenesis. *Carcinogenesis* **11**:1849–1855, 1990.

13. Russo IH, Koszalka MS and Russo J. Protective effect of chorionic gonadotropin on DMBA-induced mammary carcinogenesis. *Br J Cancer* **62**:243–247, 1990.

14. Russo IH and Russo J. Role of hCG and inhibin in breast cancer. *Int J Oncol* **4**:297–306, 1994.

15. Russo J and Russo IH. Toward a physiological approach to breast cancer prevention. *J Cancer Epidemiol Biomark Prevention* **3**:353–364, 1994.

16. Russo J, Gusterson BA, Rogers AE, Russo IH, Wellings SR and Van Zwieten MJ. Comparative study of human and rat mammary tumorigenesis. *Lab Invest* **62**:1–32, 1990.

Suggested Readings

Alexander R. McN. Finding purpose in life. *Science* 281:927, August 17, 1998.

Ayers T. Science and technology leaders discuss innovations for the future. *Science* 286:1753–1754, November 1999.

Balter M. Top scientists lock horns in research reform debate. *Science* 284:1898, June 1999.

Banville J. Beauty, charm, and strangers: science as metaphor. *Science* 281:37–39, July 3, 1998.

Becker BJ. Celestial spectroscopy: Making reality fit the myth. *Science* 301:1332–1336, September 5, 2003.

Bignami GF. Vision of the lynxes. *Science* 300:743, May 2, 2003.

Bowman JE. Tuskegee as a metaphor. *Science* 285:47, July 1999.

Bunk S. Curiosity and the scientific method. *The Scientist*: 12, April 3, 2000.

Cavalieri E, Chakravarti D, Guttenplan J, Hart E, Ingle J, Jankowiak R. Muti P, Rogan E, Russo J, Santen R and Sutter T. *Biochimica et Biophysica Acta* 1766:63–78, 2006.

Chen J-Q, Contreras RG, Wang R, Fernandez SV, Shoshani L, Russo IH, Cereijido M and Russo J. *Breast Cancer Res Treat* 96(1):1–15, 2006.

Chen J-Q, Yager JD and Russo J. *Biochimica et Biophysica Acta* XX:1–17, 2005.

Conn PM. Make science relevant, human, and clear. *The Scientist*, July 20, 1998.

Enserink M. NIH proposes rules for material exchange. *Science* 284:1445, May 1999.

Fitzpatrick SM. What makes science news newsworthy? *The Scientist*: 12, November 22, 1999.

Gentleman of science. *Science* 279:179, January 9, 1998.

Glanz J. Which way to the big bang? *Science* 284:1448–1452, May 1999.

Gould S. Deconstructing the "science wars" by reconstructing an old mold. *Science* 287:253–261, January 2000.

Gross AG and Joseph EH. What's right about science writing? *The Scientist*: 3, December 6, 1999.

Henderson A. Information science versus science policy. *Science* 289:243–248, July 14, 2000.

Holden C. Subjecting belief to the scientific method. *Science* 284:1257–1260, May 1999.

Horgan J. Science triumphant? Not so fast. *The New York Times*, January 19, 1998.

Hughes R. Silent mysteries. *Time*: 58–59, July 31, 1998.

Kennedy D. Science and secrecy. *Science* 289:724, August 2000.

Knight DM. The "great secret" of chemistry's past. *Nature* 394:633–634, August 1998.

Kreeger KY. Winning, managing, and renewing grants. *The Scientist*, October 30, 2000.

Lewis R. The anatomy of a press release. *The Scientist*: 8–12, October 26, 1998.

Lewis R. Writing book chapters broadens the scientific experience. *The Scientist*:19, June 22, 1998.

Lindee MS. Watson's world. *Science* 300:432–434, April 18, 2003.

Marias J. Palabras peligrosas. *ABC*, Jueves 2000.

Marshall E. In the crossfire: Collins on genomes, patents, and rivalry. *Science* 287:2396–2402, March 2000.

Nunney L. Are we selfish, are we nice, or are nice because we are selfish? *Science* 281, September 11, 1998.

Quittner J. Invasion of privacy. *Time*, August 25, 1997.

Russo IH and Russo J. *Eur J Canc Prev* 2:101–111, 1993.

Russo IH and Russo J. Cancer. *Women Health* 4(1):1–5, 2008.

Russo IH and Russo J. *Environ Health Perspect* 104:938–967, 1996.

Russo J and Russo IH. *J Cellular Biochem* 34:1–6, 2000.

Russo J, Lareef MH, Balogh G, Guo S and Russo IH. *J Steroid Biochem Mol Biol* 87:1–25, 2003.

Russo J and Russo IH. *Lab Invest* 57:112–137,1987.

Russo J and Russo IH. *Med Hypotheses Res* 1:11–22, 2004.

Russo J and Russo IH. *J Steroid Biochem Mol Biol* 102:89–96, 2006.

Russo J and Russo IH. *Cancer Lett* 90:81–89, 1995.

Russo J and Russo IH. The pathology of breast cancer: Staging and prognostic indicators. *JAMWA* 47:181–187, 1992.

Russo J, Hu YF, Yang X, Huang Y, Silva I, Bove B, Higgy N and Russo IH. *Front Biosci* 3:944–960, 1998.

Russo J, Hu Y-F, Silva IDCG and Russo IH. *Microsc Res Tech* 52:204–223, 2001.

Russo J, Han HJ, Kohwi Y and Kohwi-Shigematsu T. New advances in breast cancer metastasis. *Women's Health* 4(6):1–3, 2008.

Russo J and Russo IH. *Trends Endocrinol Metab* 15(5):211–214, 2004.

Russo J, Tay LK and Russo IH. *Breast Cancer Res Treat* 2:5–73, 1982.

Quittner J. Invasion of privacy. *Time*: 29–35, August 25, 1997.

Sheed W. Setting the standards. *Time*, October 5, 1998.

Stipp D. Biotech's billion dollar breakthrough. *Fortune*: 96–101, May 26, 2003.

Temple LKF, McLeod RS, Gallingers S and Wright JG. Defining disease in the genomics era. *Science* 293:807–810, August 2001.

Vogel G. TV fame and RNA glory. *Science* 301:1311–1312, September 5, 2003.

Chapter 3

The Paper

Introduction

There are numerous ways to communicate the results of the research endeavor such as scientific meetings, conferences, in review articles and most importantly, in peer-reviewed publications. Scientific meetings are opportunities for open communication with a specific audience, or sometimes to the public at large, through posters and oral presentations. To simply lecture about your work however, is not enough. The first systematically published research results came to light in France, in 1665, and since then publication in scientific journals has been the most effective way to communicate results. Scientific papers come in different formats and lengths and have different meanings and purposes. For example, *letters* are short descriptions of important current research findings that are usually fast-tracked for immediate publication because they are considered urgent. Instead, *research notes* are short descriptions of current research findings considered less urgent or important than those that appear in letters. Although letters and the research notes can be useful in establishing the importance of the results, each article must carry its own weight. Articles should contain complete descriptions of current original research findings, and range between five and 20 pages in length.

Review articles do not cover original research, rather they gather the results featured in many different articles on a particular topic into a coherent narrative about the state of affairs in that field. Review articles provide information about the topic, and also provide journal references to the original research.

The most highly regarded form of scientific communication is known in scientific parlance simply as *The Paper*, which is an article describing original research which must then be submitted to a peer-reviewed journal for publication. The apprentice of science will discover that the paper will become its own all-important entity. The paper is the end product of the research endeavor and the final product of a creative process. It is the ultimate expression of a scientist's contribution to the pool of knowledge. For as long as we keep records of human activity, the paper will surpass the existence of the individual who created it.

The nature of a peer-reviewed publication requires originality, defined as the contribution of knowledge that did not previously exist. A paper published in a *peer*-reviewed journal is also an indication that the scientist's peers have approved and accepted the new intellectual endeavor as part of common knowledge. For this process to be of true value, for it to have a lasting and legitimate impact on human understanding, it requires that the paper be executed with the utmost critical rigor, and written in an obsessively coherent manner. A manuscript not properly presented is a reflection of a careless and sloppy investigator. In addition, the manuscript must respect and acknowledge the work of others, without being a textual or conceptual copy of that work, and follow as closely as possible what the original author(s) have interpreted. Maintaining this standard is what makes the publication of data so significant, and while there is a lot of mediocre material published, this does not grant the Scientific Apprentice license to do the same.

Sloppy scientific writing is a serious problem. It reflects indifference, or worse, a belief that a journal's editors will either not detect the errors, or correct them for you. Some scientists have a magical idea that once a manuscript is mailed out, it will either be accepted, or that it is the job of the critiquing reviewers to help the author resubmit the paper to another journal. This reasoning is dangerous in that it indicates a lack of critical thinking, a trait required in order to find out what is or could be wrong with the paper. The belief that others will mend your mistakes also indicates a dependence on an outside standard of excellence rather than your own. Many papers are retracted due to fraudulence because of a lack of critical analysis of the data.

Selecting the Journal

Selecting a journal for manuscript submission is not easy, and for some, a source of frustration. In a 2008 article in *Science*, Francis S. Colins, writing about the pioneering medical geneticist Victor A. McKusick, tells the story of how as a medical student he and his brother Robert wrote a paper on the inheritance pattern of coat color in Jersey cattle, describing how the apparently dominant fawn color was actually a recessive trait. They eagerly sent their genetics paper off to the *Journal of Heredity*, but never got a response. The editor probably did not find enough merit in the paper to even muster an answer. This rejection did not, however, discourage him from a career in science.[1]

The body of scientific literature is large and there are anywhere from 10,000 to 40,000 research journals in existence. However, as Eugene Garfield notes, just a small fraction of "core" journals account for the majority of citations. And according to Garfield, about half of all articles published in 1994 came from only 500 journals, and account for more than 70 percent of what is cited worldwide. Additionally, 85 percent of published articles and 95 percent of cited articles originate from just 2,000 of the many thousands of journals. It seems that this trend in the general scientific literature also holds true for specialty and subspecialty literature. Thus, while it is impossible to read everything of relevance published on most topics, the diligent and resourceful researcher can readily follow the comparatively small core of significant literature.[2]

One thing to consider when choosing a journal to submit to is its *impact factor*. The journal impact factor, as described by Thomson Reuters, is a measure of the frequency with which the average article in a journal has been cited in a particular year. For example, the impact factor of a journal in 2008 is defined by the number of citations to all articles published in that journal from the years 2007 and 2006, divided by the number of articles published in 2007 and 2006. Eugene Garfield introduced the concept of impact factors with the idea of ranking a paper's citation frequency, but made it clear that it was not an appropriate tool for ranking individual scientists. However, Colquhoun[3] has argued that the citation rate of individual papers does not necessarily correlate to the journal's impact factor.

This disconnect is caused by skewed distribution of citation rates, meaning that high-impact journals get most of their citations from a few articles. On the other hand, Simons[4] raises another important issue: thousands of papers are published every year around the world and the quality of these papers needs to be evaluated in some objective way, not only to determine the accuracy of their contribution to fields of research, but also to help make informed decisions about rewarding scientists with funding and appointments to research positions.[4]

Impact factor is perpetuated because the number of papers published in *Science, Nature* or *Cells*, for example, is an easy way to rank the importance of a scientific journal. These journals are top-ranked journals in the field of biology with impact factors of 35 to 40 citations per article. The root of the problem surrounding impact factors is that they were created to rank individual papers, not to assess scientists and institutions. Kay Simons suggests that governments are misusing this tool to rank universities and research institutions. Hiring, faculty-promoting, and grant-awarding committees can use a journal's impact factor as a convenient shortcut to rate a paper without reading it.[4] The impact factor mentality is so prevalent that for many the plan is to send a paper to a high-rate journal, and if not accepted to continue submitting to journals in descending impact until it is finally published. This practice wastes time for editors and those who peer-review the papers. The best way to evaluate a publication is not by its impact factor but by the quality of its content. Regardless of anything else, in the end it is the quality of the work, wherever it is published, that will be judged by the scientific community.

The best historical example is Gregor Mendel's seminal publication on heredity, which was ignored for more than 30 years. When finally discovered by the scientific community, it changed our understanding of the world by establishing the basis of inheritance. An important question to consider when thinking about publication in our own time is whether Mendel was ignored because he didn't publish in a well-circulated journal or because the scientific community simply did not understand the content and significance of his paper. In his case it was likely a combination of factors.

The scientific apprentice experiences very early fear of their paper getting lost in the sea of publications and their work remaining unrecognized until published in a well-read or high-impact journal. David Colquhoun sees a silver lining for the scientific apprentice, however: "the current obsession with journal impact factors, especially at a time when, because of the Web, it matters less than ever before where an article is published."[3] While this is true, I strongly advise the scientific apprentice to develop a sense for the value and originality of their data in the way that they have interpreted it, regardless of the impact factor of the journal. In other words, impact factor should not determine the quality and depth of the work; only the scientist can do that.

A greater limiting factor for the scientific apprentice is that while more and more papers have a presence on the web, they may not be able to fully access them if the scientific apprentice's institution is not subscribed to certain journals. In periods of economic constraint this could prove to be a real barrier to the full body of extant literature and other published material. In most cases, the publishing company with the rights to the article requires a fee for granting access to full articles, and on some occasions these fees can be quite steep and increase quickly. This economic issue will skew the number of publications that will appear in journals that do not offer free access to their material over the Internet.

Authorships

There comes a point in every apprentice's life when authorship becomes a great concern because the number of papers one has authored is a measure of the productivity and continuity of one's work. Authorship and the order in which a paper is authored is a reflection of involvement in the publication. The most important decision to make is determining who will be the first author. The first author, the person whose name literally goes first under the title and is primarily linked to the paper, is also the person who has performed most of the work, put the results and interpretation together and written the manuscript. If two authors are equally responsible, it needs to be stated in a footnote that both the first and second authors are equally involved in the work. Although the first author is responsible for the paper overall, the final burden rests on the shoulders of the *senior author,* who in most cases is the principal investigator on the grant from which the funds were derived for conducting the research on which the publication is based. The senior author is also the *correspondent author,* meaning the one who receives the galley proofs and maintains contact with the editor. On some occasions the first author is the senior and correspondent author all at once. The position of the senior author, or correspondent author, in general is placed last and also stated in a footnote. If multi-authorship is taking place, which is quite common these days, the authors must be ordered according to the percentage of work respectively performed, as well as their involvement in the writing of the manuscript. This could be decided before hand or when manuscript is written. When the work is equally distributed among several members of the group, listing names in alphabetical order provides equanimity to the process.

There is no doubt that quantifying coauthor contribution is not easy. Several decades ago it was unusual to see three coauthors as it was believed that "multiple authorship endangers the author credit system,"[5] but by 2006, over 100 published papers featured over 500 coauthors, therefore, Sekercioglu[5] proposed facilitating the quantification of each author's role on the paper in relation to the contribution provided by the first author. In my opinion all the authors have merit and responsibility over the content of the publication, but the first and the senior authors are the ones who carry the bulk of responsibility over the content of the publication and who decide how to distribute the order of authorship based in the involvement of each.

Writing the Paper

The writing of a scientific paper is not an easy task and has been considered by some scientists, as Karen Young Kreeger points out, the most excruciating part of their jobs.[6] Those who enjoy putting words together might say it is a joyful act, or at least not incredibly painful. In my opinion, writing the paper, like writing a grant proposal, is one of the most challenging, rewarding and highly creative endeavors of the research scientist, independent of the experiment's outcome. For this reason I believe it is important for the scientific apprentice to have some practical guidelines at hand on how to begin writing their first paper.

Each journal has specific instructions that are readily available to authors and it is wise that they be read and understood before beginning the paper. Whereas there are some formatting differences between journals, most of the manuscripts submitted for publication consist of eight parts: title, abstract, introduction, materials and methods, results, discussion, acknowledgments, and references. Tables and figures that support the data also accompany the majority of papers. Some papers include drawings that may help to emphasize mechanisms, molecular pathways or a hypothesis that assists the reader in understanding the meaning of the results.

Materials and Methods

When writing a paper I never start at the beginning, instead I begin in the middle with the *Materials and Methods* section. It seems fitting to begin writing a paper at the same place the experiment starts; without the products of bench work there would be no results to write about. This specific portion of the manuscript is one that needs to be carefully crafted, because this is where the techniques and methodology of an experiment are explained in great detail. The *Materials and Methods* section consists of a thorough description of materials used, such as specific cell lines, animal breeds, as well as drugs and reagents, including the manufacturing and supplier information. If something like cell lines are not commercially available, provide the name of the scientists who supplied them. Seldom is this information omitted and when the reviewers do see an omission there are a series of queries for the author(s), but having to ask may give the reviewer a bad impression. This part of the *Materials and Methods* section is extremely useful to be spelled out from the beginning. It is better to start with a detailed methodology and trim it down later, if needed.

A good description of the experimental protocol is necessary, and if applicable, should include a drawing detailing the experimental scheme. For example, in our group's publication in *Cancer Research* (*Cancer Res*

67:11147–11157, 2007) we used a drawing to clarify a description of how the cells under study were collected:

> **Affymetrix microarray expression and genotyping assay.** MCF-10F cells at three different passages, designated MCF-10F1, MCF-10F2, and MCF-10F3 (passages 135, 137, and 138, respectively); trMCF cells at three different passages, designated trMCF1, trMCF2, and trMCF3 (passages 20, 22, and 23, respectively); bcMCF clones 1, 2, and 3; and caMCF 1, 2, and 3 cells were used for microarray expression and genotyping assay (Fig. 1). Total RNA and high molecular weight genomic DNA were isolated using Stat-60 (Tel-Test, Inc.) and the DNA isolation protocol previously described .....

It is vital that this section maintain adequate references that are properly cited when the techniques employed have already been published. If any technique is modified by the author(s) this needs to be stated along with an

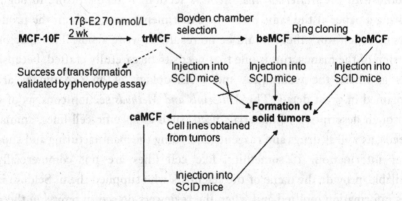

Fig. 1. Transformation of MCF-10F cells by E_2 treatment. Experimental protocol: MCF-10F cells treated with 70 nmol/L E_2 that expressed high colony efficiency and loss of ductulogenic capacity in collagen-matrix were classified as transformed (*trMCF*). Transformed cells that were invasive in a Matrigel Boyden-type invasion chambers were selected (*bsMCF*) and plated at low density for cloning (*bcMCF*). MCF-10F, trMCF, bsMCF, and bcMCF were tested for carcinogenicity by injecting them into the mammary fat pad of 45-day-old female SCID mice. MCF-10F and trMCF cells did not induce tumors (*canceled arrow*); bsMCF and bcMCF formed solid tumors from which four cell lines, identified as caMCF, were derived and proven to be tumorigenic in SCID mice.[7]

explanation of the changes. The procedures followed, as well as the reagents used, need to be clearly described; if the data being published are interesting and novel, other scientists will be eager to repeat the study, maybe using other cells or different models or in other areas of focus, but with the same methodological approach. If statistical tools have been used, the type of analysis and the power calculation must be clearly stated. If human material is used in the work, how it was obtained must be explained and if adequate, approval from the Internal Review Board (IRB) must be indicated. The role of the IRB at each institution is to maintain an effective Human Research Protection Program to protect the rights and welfare of human participants. If animals are used, adequate certification from the Institutional Animal Care and Use Committee (IACUC) must also be provided. The IACUC is a self-regulating entity that, according the US federal law, must be established by institutions that use laboratory animals for research or instructional purposes to oversee and evaluate all aspects of the institution animal care and use program.

A thorough description of the materials and methods provides an important reference for the reviewer and indicates the quality of the research performed and mastery of the methodology on the part of the scientific apprentice. For example, when gene analysis is performed, as in the case of microarrays (*Cancer Res* **67**:11147–11157, 2007), the platform and suppliers must be identified, as well as the software utilized for data analysis.

Differentially expressed genes. The functional profiles were represented by the biological processes in the Gene Ontology (GO) database. The number of dysregulated genes in each chromosome or GO category was compared with that of all genes in the HG-U133_Plus_2 chip to determine the significance of the chromosome or GO category. The analysis was performed using Onto-Express, with the default selection of statistical method (hypergeometric distribution followed by false discovery rate correction). The three lists of genes dysregulated in trMCF, bcMCF, or caMCF were uploaded into Onto-Express to identify significant GO categories (q $\leq$ 0.05 with five or more genes). The up- or down-regulated genes were uploaded into Onto-Express separately to identify the individual chromosomes enriched (q $\leq$ 0.0001) with these genes.

The above paragraph shows that the author fully understands every aspect of the work and is actively interested in repeatability.

Results

Next comes the *Results* section. Whereas every set of data requires a different kind of description, it is useful to go from the general to the specific. For example, if animals were used it is better to explain whether the treatment produced any changes in the general state of the animals, such as weight. If tumors were induced, clearly state how many tumors were obtained and provide tables and graphs documenting the finding. If tumors were found, the histological type should be described, in addition to the type of classification, referencing the materials and methods when applicable. If cell cultures were a component of the experiment, first describe the phenotypes of the cells being studied, and only then go on to describe the main biochemical finding or molecular alterations depicted. If morphological changes were observed and documented by special techniques such as fluorescence, confocal or electron microscopy, be sure that the description matches the images. Moving into specific details step by step ensures that61the reader will easily follow your description of the experiment and therefore better grasp its implications.

Never let the reader fend for themselves to understand tables or figures. Succinctly and clearly describe your data in the body of the text and use figures to clarify your description. If needed, give additional information in the figure's legend or in tables. The following paragraph describes the results reported in a paper about MCF10F cells transformed by estradiol written by Huang *et al.* and published in *Cancer Research* in 2008:[7]

Analysis of chromosome copy number and LOH in neoplastically transformed MCF-10F cells. MCF-10F cells that after treatment with E_2 expressed high colony efficiency and loss of ductulogenic capacity in collagen-matrix represented the first level of *in vitro* transformation. Cells expressing these two variables were classified as transformed (trMCF), which after further selection for invasiveness in a Matrigel invasion chamber originated the second level of transformation: the invasive (bsMCF) and the cloned (bcMCF) cells (Fig. 1). The bsMCF cells formed tumors in SCID mice from which four cell lines, caMCF, were derived (Fig. 1). By ring cloning, seven subclones were isolated from the invasive bsMCF cells: bcMCF-1, bcMCF-2, bcMCF-3, bcMCF-4, bcMCF-5, bcMCF-6, and bcMCF-7. All the bcMCF subclones produced invasive poorly differentiated tumors in SCID mice with different morphologic phenotypes: spindle cell type (bcMCF-1 and bcMCF-4), epithelial cell type (bcMCF-2, bcMCF-6, and bcMCF-7), and with mix features of spindle and epithelial type. As it was previously reported, MCF-10F cells were seeded on Boyden chamber as control; cells that passed through the membrane were selected, expanded, and injected in SCID mice; these cells did not produce tumors.

The above paragraph succinctly describes how the different cell lines were obtained. It explains, for example, how the different subclones of cells were obtained and how they produced tumors when injected in SCID mice. The importance of a detailed description is that the results must match what was described in the *Materials and Methods* and the reader should be able to easily find the procedures that lead to the described results. At the end of each section provide a small summary, one or two sentences, that capture the essence of the data described. Do not forget that while your paper is of utmost importance to you, it is one of the hundreds and thousands that any given reader, be it a reviewer or a fellow scientist, will read. **The clearer the message, the higher the impact.**

While compiling the results it is possible to realize that you have come upon highly relevant, stimulating data that are so novel they could change the present paradigm of a biological process. If this is the case, the data need to be critically evaluated and the results of each data piece confirmed using more than one approach to be sure that the findings are consistent. In this regard data analysis and the writing of the results are a commingled operation.

But what happens when your results give you the opposite of what you hoped, or yield a negative result? Whereas the first impulse is to disregard them, this is the wrong attitude. If the experiments have been done and analyzed properly, that simply means that the results are telling us something unexpected, and perhaps even more meaningful than what we had imagined. It is true that scientists aim to publish the results of successful experiments, and are generally less excited about trumpeting those that simply confirm a null hypothesis, for example, that a particular genetic marker is not associated with an inherited disease, or that there is no difference between mice given a candidate drug and those in the control group. In his 2003 article "Null and Void,"[8,9] Jonathan Knight addresses the fact that most researchers do not bother to write up negative results. I will advise the scientific apprentice to be cautious in publishing negative results, however. Negative data must be critically evaluated in order to contribute to the overall scheme of the present knowledge. It might be the case that a paper containing data that confirms the null hypothesis "convincingly overthrows a widely held belief."[8]

To stress the importance of clarity, Knight reminds his readers of Francis Crick's somewhat exasperated observation, "There is no form of prose more difficult to understand and more tedious to read than the average scientific paper." Knight continues, "Crick and others of his generation, who began writing scientific papers in the 1940s, have witnessed the transformation of scientific prose. A form that was as readable as the average newspaper has, in some fields, become a jungle of jargon that even those familiar with the territory struggle to understand."[8,9] This difficulty in expressing complex processes in simple words is in part due to the lexicon created by the numerous disciplines and new findings in fields such as molecular biology and immunology. This language is not easy to replace, and necessarily, cannot be replaced. Nevertheless, when concepts demand a specialized vocabulary, clarification is needed to ensure that their true meaning remains apparent. Most likely this does not require special language degrees, rather, a logical thinking process that allows complexity to be unfolded in easily readable and understandable prose by the average reader. Language experts generally agree that a better measure of accessibility is whether a piece of writing contains words in common usage — those that are at the front of the reader's mind, rather than those tucked away in the recesses of memory. No doubt it will always be necessary to use terms specialized to a particular field. However, by not using unnecessarily complex or circuitous language, the complex *ideas* of the paper will be all the more accessible.

Writing the *Results* requires an objective mind. In describing what has been created from our studies, we must avoid releasing other people's data unless the scientists involved agree. In this era of instant access, this issue has acquired grave importance. The Internet has allowed knowledge to be easily transmitted, and out of this abundant source of data, opinion, and miscellanea, the boundaries of intellectual property have often been broken. Daniel Frank clearly discusses this issue and poses the question: Is information released in a fair and representative manner?[10] Trustworthiness regarding the dissemination of findings is a rule that the scientific apprentice must embed into his or her being. Breaking this rule by not crediting an original source of information, as Frank indicates, will violate "the spirit of collegiality that most scientists hold as an ideal in our public discourse. We all accept that others may scoop our work. We should not have to worry about being scooped by our own data"[10]

Preparation of figures and tables

Figures and tables are the visual manifestation of research data. This has always been very obvious to me because in my early years of my training I did a great deal of electron microscopic studies. For me at that time, obtaining the right photo at the correct magnification with optimal contrast was just the beginning of preparing a paper. This involved many hours in the darkroom that instilled great respect and appreciation for photography and is something that I have continued to nurture to this very day. This early experience makes me especially sensitive to the preparation of figures and tables for publication. My first piece of advice to the scientific apprentice is a practical one: read the graphics specifications issued by the journal to which you are planning to submit your work before you start preparing your figures. As I was preparing to write this portion of the chapter I happened to find a very enlightening publication written by Liana Holmberg[11] that reviews these concepts very clearly. Holmberg stresses the importance of looking for the resolution requirements for each type of graphic, and preferred file formats, in order to minimize unpleasant, last minute surprises. As opposed to earlier years, most graphic and photo materials are digitized and can be easily edited using readily available software like Adobe Photoshop. But even though imaging processes have gotten faster and less strenuous, it is important to know how to master them in order to provide a journal with images at the adequate resolution and in the proper format.[11] It should be understood by the scientific apprentice that while all the materials generated into tables and figures should already be documented in the Laboratory Protocol Books (which will be discussed in more detail in Chapter 7), it is absolutely necessary to keep the original material in a safe place in case the material submitted for publication is lost in the mail or, as happens less frequently, is misplaced by the editor. In addition, this material might be needed for a review article, or a book even though it may have already been copyrighted. If you are using digital material it is important to save it in a high-resolution original version and be rigorous about backing up your files.

It may seem that keeping track of where and when your graphic material is published is a minor concern, but once you begin publishing frequently, the chances of submitting the same figure twice increases, especially if you publish several review articles on the same subject. Therefore, keep records and track your tables and figures because they are not only important for your publications, but also for your oral presentations. If your resulting material is of good visual quality, other investigators will ask whether they can use it for their own presentations. In this case it is important that you develop a trademark to avoid plagiarism or misuse of your graphic material. I mention this because one of the speakers at a conference I attended showed one of my drawings from one of my earlier papers in his presentation. I was excited to see my work presented by others; however, this excitement turned to bewilderment when the speaker not only failed to recognize the original author of the work, but went so far as to refer to it as his own. It is simultaneously comforting and painful when you see your work being referenced, yet know that your labors are not being acknowledged. Therefore, always be sure that your work is identified and be extremely respectful when using material that belongs to others, either in oral presentations or written work. Although the journals are vigilant about copyright issues, it can be easier to overlook these issues in oral presentations.

Introduction

Once the material and methods and the results have been written, it is then time to write the *Introduction*. The *Introduction* should be brief but clearly state the problem that the research is addressing. For example, it is well known that various types of contaminants, such as butyl benzyl phthalate, or BBP, are extremely prevalent in our environment. When we chose to study this particular compound's effect on the genomic signature of the mammary gland, our ultimate aim was to understand how something so common could be carcinogenic. Experiments are not designed on whimsy; they are put in service to fulfill a need for understanding. The introduction's purpose is to orient the reader to the greater context of your idea.

The introduction should also provide a historical perspective on how the subject has been studied and what is already known. This background provides the basis for explaining the objective of the present work, demonstrates that the author has approached the subject in an innovative way, and outlines what will be presented in the rest of the paper. The paragraph below is the introduction of a paper published by our group:[7]

> Breast cancer is a malignancy whose dependence on ovarian function was shown by Beatson,[12] who induced regression of advanced cancer in premenopausal women by surgically removing the ovaries.[7]

Here we acknowledge the problem and the investigator who first observed the link between estrogen and cancer. The following passage refers to some seminal work done after Beatson's first observation:

> Thereafter, the same procedure was also proven to control the progression of metastatic disease.[13] The identification of estrogen (E_2) production by the ovaries, the isolation of the estrogen receptor (ER) protein, and the

greater incidence of ERα-positive tumors observed in postmenopausal women led to the identification of a strong association between estrogen exposure with increased breast cancer risk.[14]

The general framework of the manuscript is established by laying out what is already known and what needs to be done in order to understand the mechanism of action of estrogen and cancer:

Despite the epidemiologic and clinical evidence linking cumulative and sustained exposure to estrogens with increased risk of developing breast cancer[14] the ultimate mechanisms by which estrogens induce cancer and the specific cells they act upon for initiating malignant transformation have not been fully identified. Among the mechanisms of estrogen action, the most widely acknowledged is the binding of the hormone to its specific nuclear ERα, initiating a signal that is potently mitogenic.[15]

In the next paragraph, the author must establish a critical point to the accepted concepts by bringing additional references.

However, the fact that ERα knockout mice expressing the Wnt-1 oncogene (ERKO/Wnt-1) develop mammary tumors in response to treatment with estrogen provides direct evidence that E$_2$ may cause breast cancer through a genotoxic, non-ERα–mediated mechanism.[16] This postulate is further supported by the observation that when ovariectomized mice are supplemented with estrogen, they develop a higher tumor incidence with shorter latency time than control animals, even in the presence of the pure antiestrogen ICI-182,780. Experimental studies on estrogen metabolism,[17] formation of DNA adducts (18), carcinogenicity,[18,19] mutagenicity,[20] and cell transformation[21,22] have supported the hypothesis that reaction of specific estrogen metabolites, namely catechol estrogen-3,4-quinones (CE-3,4-Q) and, to a much lesser extent, CE-2,3-Q, can generate critical DNA mutations that initiate breast, prostate, and other cancers.[23–25]

We further narrow the information on how the *in vitro* model was used in our study, and discuss previous publications which support the idea that the *in vitro* model mimics the disease in humans.

Our observations that ductal carcinomas originate in lobules type 1 (Lob.1) of the immature breast,[26] which are the structures with the highest proliferative activity and highest percentage of ERα and progesterone receptor–positive cells, provide a mechanistic explanation for the higher susceptibility of these structures to undergo neoplastic transformation when exposed to chemical carcinogens, as shown by *in vitro* experiments.[27] However, the role of ERα-positive and ERα-negative cells in the initiation of breast cancer is not clear. The fact that the cells that do proliferate in culture are ERα-negative suggests that the stem cells that originate cancer are the ERα-negative proliferating cells. This idea is further supported by our observations that MCF-10F, a spontaneously immortalized ERα-negative human breast epithelial cell line derived from breast tissues containing Lob.1 and Lob.2,[28] becomes malignant after exposure to the chemical carcinogen benz(a)pyrene[27] and 17β-estradiol (E$_2$; refs. 22, 29). Breast cancer has been subdivided into five major subtypes: basal-like, Her2 (ERBB2)–overexpressing, normal breast tissue-like, and two subtypes of luminal-like, luminal A and luminal B.[30] The luminal-like subtypes display moderate to high expression of ERα and luminal cytokeratins, whereas the basal-like subtype is negative for both ERα and ERBB2, with high expression of basal cytokeratins 5 and 17. The ERBB2-overexpressing subtype is also ERα negative and, like the basal-like tumors, is associated with poorer prognosis as measured as time to development of distal metastasis.[30–32] Altogether, these data support the concept that ER-positive and ER-negative tumors may originate from two different cell populations, as postulated earlier.[33,34] In addition to differences inherent to the type of cell in which cancer originates, neoplastically initiated cells lose specific characteristics of epithelial differentiation as the result of their progression toward malignancy. As the epithelial cells lose their polarity and cell-to-cell junctions, regulated in part by the expression of E-cadherin, they acquire characteristics of mesenchymal cells, which lack stable intercellular junctions.[31] This epithelial to mesenchymal transition (EMT) leads to exacerbation of motility and invasiveness in many cell types and is often considered a prerequisite for tumor infiltration and metastasis.[31]

In the final paragraph of the *Introduction*, we stated the specific layout of the manuscript and provided a preview of the data that would follow.

To outline the pathways through which estrogen acts as carcinogen in the human breast (i.e., either through the receptor pathway or through a genotoxic effect in a specific cell type of the breast), we used an *in vitro–in vivo* system in which the spontaneously immortalized ERα-negative human breast epithelial cell (HBEC) line MCF-10F was transformed by treatment with E_2.[33] E_2-transformed cells progressively express phenotypes of *in vitro* cell transformation, including colony formation in agar methocel, decreased ductulogenesis, increased invasiveness in a Matrigel invasion system, and tumorigenesis in a heterologous host. Tumors formed in severe combined immunodeficient (SCID) mice by invasive cells and by cell lines derived from those tumors were poorly differentiated ERα-, progesterone receptor–, and ERBB2-negative adenocarcinomas.[34] These characteristics are similar to those of basal cell type primary carcinomas previously described.[35] To better understand the molecular events associated with the progressive phenotypic changes that were observed during estrogen-mediated malignant cell transformation, we performed Affymetrix 100k single nucleotide polymorphism (SNP) arrays to measure chromosomal copy number and loss of heterozygosity (LOH), and HG–U133_Plus_2 array for analyzing mRNA expression in MCF-10F cells at different stages of cell transformation. By integrating these data, we were able to identify associations between copy number changes, LOH, and tumorigenic phenotype, as well as the related changes in transcript expression. Functional analyses of these data identified several dysregulated pathways associated with progressive tumorigenic and invasive capacity.

By the end of this introduction the reader must have a clear idea of the problem, the historical basis of it, and how the author's data will enhance the knowledge available up to that moment.

Discussion

The *Discussion* of the paper has three main components: a brief summary of the data to be discussed, not to exceed two to three sentences, the core, and concluding remarks.

The paragraph below summarizes the results by briefly describing their salient points, and underscores the message that the paper aims to transmit.

This study integrates structural and functional genomic data analyses to elucidate the progressive molecular events in the E2-mediated malignant transformation of ER (–) HBECs. Genomic aberrations progressively accumulated as the cells expressed more aggressive phenotypes (i.e., in the tumorigenic bcMCF and caMCF) in comparison with the nontumorigenic trMCF cells. Accordingly, the number of genes with altered levels of expression was greater in the tumorigenic cells, as where the chromosomes enriched with up- or down-regulated genes. Importantly, the 12 samples were correctly classified into tumorigenic or nontumorigenic groups based on the profile of copy number changes, indicating that in our model changes in copy number provide a genomic signature for the tumorigenic phenotype. Together, these findings revealed an intrinsic link between E2-induced copy number changes, gene expression alterations, and tumorigenesis in ERα (–) HBECs.

The core of the *Discussion* is a careful dissection of the results in light of current published data and emphasizes the new path that the results open. Here is where the scientific apprentice's scientific knowledge comes into play. For example, if the data shows that an agent has new therapeutic properties in tumor development, this finding needs to be supported by indicating whether this compound has already been discovered by another scientist, and if so, how it has been used. Perhaps the compound has existed for many years. If this is the case, the original citation must be mentioned and the latest reference of an updated review article that comprehensively describes it. A warning here however — avoid relying only on reviews. To only cite review articles is unfair to the original researcher who made the initial discovery. Even if the finding took place centuries ago and the scientist is long gone, they deserve the respect of having their work be used as a primary source. There have been times in our research endeavors that while we were the first to establish, for example, the connection between the process of gland differentiation and the protective effect against cancer that pregnancy exerts on the mammary gland, in reading other papers I have found that our work was credited to another investigator who did nothing more that write a review article. This has occurred many times and shows lack of careful data analysis.

Another common practice is searching only the latest references without paying attention to how the concept originated. I cannot emphasize this enough: be respectful to those who have generated the initial kernel of knowledge. We must learn to respect those who laid the groundwork for us today.

Once you have established the basis of the work, proceed by elaborating your discovery. The author needs to indicate that, going back to our previous example, a compound used in mammary tumorigenesis has never been used before and that the author is the first to do so. Once this has been established you can then start discussing the uniqueness of the findings and compare the effect of your compound with others. At this point the author should defend his or her position by explaining why the data are solid. This involves honestly evaluating the experimental methods, the statistical power utilized, indicating whether the noted differences are truly significant, that the model used was a well-established one, and that the manner in which the samples were collected was appropriate, the diagnosis was foolproof, and finally, indicate what measures the author used to confirm that the results were specific and not influenced by confounding factors. Thus it is important to indicate the positive and negative controls used in all experiments. By the end of this portion of the discussion the author should have established the validity and the novelty of the data. In this way they are assuring the reader that the findings are not merely coincidental but that the results are the consequence of a well-planned experiment in which all the main components have been considered.

The discussion is closed by speculating on the significance of the data. This is where the author is entitled to indicate the relevance of this data to the overall area of study, such as breast cancer. An author must carefully evaluate the words they use — he or she needs to convey to the reader that the paper has established a novel piece of information that may help others follow in corroboration with other models, or in translating this knowledge to the clinic. Whether any given paper will be a seminal piece of work that will determine the scientific apprentice's career is difficult to know, but it could be a seed for the paper that follows it.

The next paragraph is an example of concluding remarks we used in the discussion of the paper we are citing as an example.[7]

Our results support the concept that E_2-induced breast cancer is a polygenic disease having a large range of genomic instabilities. E_2 and/or its metabolites can directly cause genomic aberrations without the mediation of ERα. The genomic aberrations lead to changes in gene expression, which result in disrupted integrin signaling and apoptotic pathways, and epithelial to mesenchymal transition. These functional changes lead to colony formation in agar-methocel, loss of ductulogenesis in collagen matrix, invasiveness *in vitro*, and tumor growth in SCID mice *in vivo*.[34] However, MCF-10F is a spontaneously immortalized cell line harvested from a woman's breast that was free of malignancies, but had a diagnosis of benign fibrocystic disease.[36]

At the end of the discussion address unexplored areas that your work has opened. For example, you may have become interested in searching out similar compounds within the same family of substances of the compound you began studying, perhaps through your research you have found a need to explore molecular pathways of action. You may find it important to explore how your data translates to the clinic, as we did in a manuscript published in *Cancer Epidemiology, Biomarkers and Prevention*[43]:

Furthermore, this genomic signature is constituted by genes that cluster differently than those genes expressed in the epithelial cells of parous and nulliparous women with breast cancer as well as from nulliparous women without cancer. This genomic signature allowed us to evaluate the degree of mammary gland differentiation induced by pregnancy. Of importance is the fact that this signature serves for characterizing at molecular level the fully differentiated condition of the breast epithelium that is associated with a reduction in breast cancer risk, thus providing a useful molecular tool for predicting when pregnancy has been protective, for identifying women at risk irrespective of their pregnancy history, and for its use as an intermediate biomarker for evaluating cancer preventive agents.

On other occasions you are acknowledging the possibility that your findings may not apply in the exact same way to other types of cells or tissues[7]:

> Hence, we cannot rule out its inherited susceptibility to estrogen and the disposition to tumorigenesis. Therefore, more normal primary human breast epithelial cell lines should be studied to validate and elaborate on the molecular mechanisms unveiled by our results.[7]

The Title and the Abstract

The *Title* and the *Abstract* are the first things the reader looks at and it's likely they will be the first things found in any literature search, therefore, these two parts of the manuscript must be very well crafted.

Whereas the *Abstract* must be written when the final draft of the manuscript is finished, developing the title is a process; the title takes shape throughout the drafting of the paper. The title must be 80 to 120 characters in length and it must contain the central concept that the author(s) want to transmit. Most journals have restrictions regarding the length of the title, but as a general rule, a conceptual title is favored over a descriptive one. Write at least six different titles and from those select the one that best embodies the spirit of the work.

The following are examples of different manuscript titles:

"7,12-Dimethylbenz(a)anthracene (DMBA) induced DNA binding and repair synthesis in susceptible and non-susceptible mammary epithelial cells in culture,"[44] contains a description of the content of the manuscript, and indicates the specificity of the study. For example, the words *"DNA binding and repair synthesis,"* indicate the two main end points measured. The phrase, *"in susceptible and non-susceptible mammary epithelial cells in culture"* describe the model utilized as well as further elaborates on the cell populations in question (susceptible and non-susceptible mammary epithelial).

In contrast, *"Differentiation of the mammary gland and susceptibility to carcinogenesis"*[45] or *"Biological and molecular basis of mammary carcinogenesis"*[46] both describe fields of research in and of themselves and are better suited as titles for review articles rather than original ones. Instead *"Calcium modula-tion on microtubule assembly in human breast epithelial cells"*[47] or *"Protective effect of chorionic gonadotropin on DMBA-induced mammary carcinogenesis"*[48] are both descriptive of the experiments *and* define their fields of study.

The abstract, as the word *abstract* suggests, is the portion of the paper that tells the whole story in 200 to 300 words. In the paper published by Moral *et al.*[49] titled *"The plasticizer butyl benzyl phthalate induces genomic changes in rat mammary gland after neonatal/prepubertal exposure,"* the jour-nal requested a short abstract dividing background, results and conclusions:

Background

Phthalate esters like n-butyl benzyl phthalate (BBP) are widely used plasticizers. BBP has shown endocrine-disrupting properties, thus hav-ing a potential effect on hormone-sensitive tissues. The aim of this study is to determine the effect of neonatal/prepubertal exposure (post-natal days 2–20) to BBP on maturation parameters and on the morphology, proliferative index and genomic signature of the rat mammary gland at different ages of development (21, 35, 50 and 100 days).

Results

Here we show that exposure to BBP increased the uterine weight/body weight ratio at 21 days and decreased the body weight at time of vaginal opening. BBP did not induce significant changes on the morphology of the mammary gland, but increased proliferative index in terminal end buds at 35 days and in lobules 1 at several ages. Moreover, BBP had an effect on the genomic profile of the mammary gland mainly at the end of the exposure (21 days), becoming less prominent thereafter. By this age a significant number of genes related to proliferation and differentiation, communication and sig-nal transduction were up-regulated in the glands of the exposed animals.

Conclusions

These results suggest that BBP has an effect in the gene expression profile of the mammary gland.

In other cases, the abstract briefly summarizes the work without separating the background from the results, even though the information is still there. This is the case in an abstract published by IH Russo[50]:

Cancer of the breast is experiencing a worldwide increase in women from all ethnic backgrounds and in association with westernization. Although the agent that initiates the disease and the mechanisms that stimulate its progression are not known, lifestyle changes and environmental exposures occurring during the last decades are considered to exert a significant influence in the global increase in breast cancer incidence. Modern societies have a reduced population of women that are protected by an early pregnancy, mainly due to postponement of motherhood, reduction in family size, and increasing infertility. These changes result in a lack of breast maturation. Human chorionic gonadotropin (hCG), a hormone produced during pregnancy, stimulates breast differentiation and imprints in it a genomic signature similar to that imprinted by pregnancy. Its administration to high risk childless women might represent the most physiological approach to breast cancer prevention.[50]

Some journals, like *Nature*, also require very short abstracts, even less than 200 words, such as the one published by Hye-Jung Han *et al.*[51]:

Mechanisms underlying global changes in gene expression during tumour progression are poorly understood. SATB1 is a genome organizer that tethers multiple genomic loci and recruits chromatin-remodeling enzymes to regulate chromatin structure and gene expression. Here we show that SATB1 is expressed by aggressive breast cancer cells and its expression level has high prognostic significance ($P < 0.0001$), independent of lymph-node status. RNA-interference-mediated knockdown of SATB1 in highly aggressive (MDA-MB-231) cancer cells altered the expression of >1,000 genes, reversing tumorigenesis by restoring breast-like acinar polarity and inhibiting tumour growth and metastasis *in vivo*. Conversely, ectopic SATB1 expression in non-aggressive (SKBR3) cells led to gene expression patterns consistent with aggressive-tumour phenotypes, acquiring metastatic activity *in vivo*. SATB1 delineates specific epigenetic modifications at target gene loci, directly upregulating metastasis-associated genes while downregulating tumour-suppressor genes. SATB1 reprogrammes chromatin organization and the transcription profiles of breast tumours to promote growth and metastasis; this is a new mechanism of tumour progression.[51]

Acknowledgments

In this section of the paper you must acknowledge those who might not have directly contributed, but that without their help the final product would have been impossible. Begin by acknowledging the technical people that contributed to the collection of the data. Next, thank those with whom you consulted and gave you free advice, reviewed the manuscript and provided vital suggestions, revised the language or typed the final product, as well as the librarian who helped you search for references not easily found on the Internet. Acknowledge those who took care of the animals, the facilities that you used, or who in one way or another facilitated the gathering of the data. Importantly, never forget your mentors and the institutions that provided the funds making your research possible. This is an important parameter for the granting institutions to gauge how their money has been expended.

References

As indicated above, a journal's instructions to authors explain in detail how the journal prefers references cited and this is generally easy to follow. In addition there are many types of software dedicated to keeping track of references and formatting. It is important that the scientific apprentice be aware that references are the support that makes the data solid and valid in concert with the specific knowledge that he or she is reporting. This is an important concept that authors must bear in mind at all times. Additionally, the scientific apprentice must be honest in providing a historic perspective of the knowledge under discussion by giving adequate credit to the people who have been working in the area before them, and cite even those who you think might not agree with your findings. Do not think you can hide by omitting citations to your adversaries' work because you are afraid that they might give your work a bad review. Whether you cite them or not, the journal's editor will use those names or references and request that they revise and evaluate your paper precisely because they hold an opposing view. Therefore, the accuracy of each reference is crucial. If your data are in opposition of what has been previously described, rather than ignoring these previous studies, acknowledge them while at the same time drawing attention to those studies that do support your work. If you explain why your data differs from someone else's, they will hopefully understand your point, and even though they may not agree with your statements and conclusions, they will be obliged to give a rational opinion that will help you rewrite the discussion, or to spell out the differences better.

The Role of the Reviewers

Floyd E. Bloom stated, "regardless of the form in which manuscripts are transmitted between authors, editors, and reviewers, the integrity of the scientific review process requires that the performance of reviewers be appropriately rewarded."[37] There is no escaping the fact that the reviewing process is a painful one, yet remains the best way to evaluate the quality of research work. The strength of the system lies in the fact that generally three or four, sometimes more, experts in the field review a manuscript. They provide assurance that the published work is executed according to well-established criteria, that the data have been properly collected and analyzed, and that the conclusions and information provided will advance the field. To fulfill these requirements, the editor in charge of the journal, or the associate editors, select the reviewers and make certain critical decisions, such as determining who the author's peers are and who has the most knowledge and experience in the field. Although not difficult to find, the most worrisome part is that not all of the potentially excellent reviewers are available. The reality is that reviewers are in high demand and do not have enough time to accept every request for manuscript review. I accept only 5% of the requests I receive for manuscript revision and that 5% adds up to about five manuscripts a month, which is equivalent to approximately 20 hours of work. I agree with Bloom when he writes, "only the most dedicated scholars will agree to review frequently, given the pressures they face from research, grant proposals, and their own papers."[37]

The scientific apprentice should also know that the reviewer's task is a rewarding one because it allows them to be updated on new developments. In my view, reviewing is a way to acknowledge your peers' effort in the hopes that they might also dedicate time to reviewing your work. I will never forget the help of those who reviewed my early manuscripts, and I am grateful to them. Reviewers are important, not only for the author of the manuscript, but also for maintaining the intellectual life of researchers.

Peter A. Lawrence writes, "Scientists are increasingly desperate to publish in a few top journals and are wasting time and energy manipulating their manuscripts and courting editors. As a result, the objective presentation of work, the accessibility of articles and the quality of research itself are being compromised."[38] Lawrence's is a rather dark view, but unfortunately, an accurate one. A major problem the scientific apprentice will face in the present research environment will be the ongoing struggle between publication and the most important principle of all: the quality and integrity of the research work.

Plagiarism and Copyrights

Gretchen Vogel[39] and Declan Butler[40] in their articles published in *Science* and *Nature*, respectively, discuss plagiarism by addressing different aspects of the problem — from alteration of single images, to full replication of an entire manuscript. I strongly urge the scientific apprentice to read these two articles for themselves because they describe the magnitude of the problem and outline the consequences for both scientific apprentices, their mentors, and the institutions involved.

In order to protect knowledge generated, copyright procedure has been established by most journals. Floyd E. Bloom explains it well in his article "The Rightness of Copyright:[41]" "copyright transfer is critical to the process of communicating scientific information accurately. Neither the public nor the scientific community benefits from the potentially no-holds-barred electronic dissemination capability provided by today's Internet tools. Much information on the Internet may be free, but quality information worthy of appreciation requires more effort than most scientists could muster, even if able." In contrast, there are those like Steven Bachrach *et al.*[42] who published an article, "Who Should Own Scientific Papers?" emphasizing "the importance of electronic communication by offering many potential improvements to enhance traditional publication, scientists, administrators, and federal science policymakers must reconsider both how the results of publicly funded research are best disseminated and how that dissemination is best supported."

It is sobering to think that scientific research is not an imposed obligation but a self-chosen lifestyle. As a consequence, most of the problems that we face with plagiarism, copyright and ownership of ideas have emerged by science managers that, in their attempt to help organize the complexity of scientific research, have established rules and regulations that while helpful in some instances, incite those not genuinely committed to the research endeavor to seek easy answers, creating the need for more regulations. My intention is not to provide instruction concerning writing skills, but I know from experience that writing is a long process of trial and error. More importantly, the paper's function is to deliver a message, therefore it needs to be clear. Clear writing is the result of clear thinking, thus the scientific apprentice needs to first learn how to think clearly and logically. Better writing is the byproduct of this effort.

References

1. Collins FS and McKusick VA. *Science* **321**, August 2008.
2. Garfield E. The significant scientific literature appears in a small core of journals. *The Scientist,* September 2, 1996.
3. Colquhoun D. Challenging the tyranny of impact factors. *Nature* **423**, May 29, 2003.
4. Simons K. The misused impact factor. *Science* **322**, October 10, 2008.
5. Sekercioglu CH. Quantifying coauthor contributions. *Science* **322**, October 17, 2008.

6. Kreeger KY. Writing Science. *The Scientist*, January 10, 2000.

7. Huang Y, Fernandez S, Goodwin S, Russo PA, Russo IH, Sutter T and Russo J. Epithelial to mesenchymal transition in human breast epithelial cells transformed by 17- beta- estradiol. *Cancer Res* 67:11147–11157, 2007.

8. Knight J. Null and void. *Nature*, April 10, 2003.

9. Knight J. Clear as mud. *Nature*, May 22, 2003.

10. Frank DN. Don't release other people's data without their consent. *Nature* 455, October 2, 2008.

11. Holmberg L. What happened to my figures? *The ASCB Newsletter* 27(8) August 2004.

12. Beatson G. On the treatment of inoperable cases of carcinoma of the mammary. Suggestions for new method of treatment with illustrative cases. *Lancet* 2:104–107, 1896.

13. Boyd S. An oophorectomy in cancer of the breast. *Br Med J* 2:1161–1167, 1900.

14. Henderson BE, Ross R and Bernstein L. Estrogens as a cause of human cancer: The Richard and Hinda Rosenthal Foundation award lecture. *Cancer Res* 48:246–253, 1988.

15. Suga S, Kato K, Ohgami T *et al.* An inhibitory effect on cell proliferation by blockage of the MAPK/estrogen receptor/MDM2 signal pathway in gynecologic cancer. *Gynecol Oncol* 105:341–350, 2007.

16. Bocchinfuso WP, Hively WP, Couse JF, Varmus HE and Korach KS. A mouse mammary tumor virus-Wnt-1 transgene induces mammary gland hyperplasia and tumorigenesis in mice lacking estrogen receptor-α. *Cancer Res* 59:1869–1876, 1999.

17. Rogan EG, Badawi AF, Devanesan PD *et al.* Relative imbalances in estrogen metabolism and conjugation in breast tissue of women with carcinoma: Potential biomarkers of susceptibility to cancer. *Carcinogenesis* 24:697–702, 2003.

18. Cavalieri EL, Stack DE, Devanesan PD *et al.* Molecular origin of cancer: Catechol estrogen-3,4-quinones as endogenous tumor initiators. *Proc Natl Acad Sci USA* 94:10937–10942, 1997.

19. Li JJ and Li SA. Estrogen carcinogenesis in Syrian hamster tissues: Role of metabolism. *Fed Proc* 46:1858–1863, 1987.

20. Newbold RR and Liehr JG. Induction of uterine adenocarcinoma in CD-1 mice by catechol estrogens. *Cancer Res* 60:235–237, 2000.

21. Chakravarti D, Mailander PC, Li KM *et al.* Evidence that a burst of DNA depurination in SENCAR mouse skin induces error-prone repair and forms mutations in the H-ras gene. *Oncogene* **20**:7945–7953, 2001.

22. Fernandez SV, Russo IH, Lareef M, Balsara B and Russo J. Comparative genomic hybridization of human breast epithelial cells transformed by estrogen and its metabolites. *Int J Oncol* **26**:691–695, 2005.

23. Lareef MH, Garber J, Russo PA *et al.* The estrogen antagonist ICI-182–780 does not inhibit the transformation phenotypes induced by 17-β-estradiol and 4-OH estradiol in human breast epithelial cells. *Int J Oncol* **26**:423–429, 2005.

24. Cavalieri E, Rogan E and Chakravarti D. The role of endogenous catechol quinones in the initiation of cancer and neurodegenerative diseases. *Methods Enzymol* **382**:293–319, 2004.

25. Russo J, Hasan Lareef M, Balogh G, Guo S and Russo IH. Estrogen and its metabolites are carcinogenic agents in human breast epithelial cells. *J Steroid Biochem Mol Biol* **87**:1–25, 2003.

26. Russo J, Gusterson BA, Rogers AE *et al.* Comparative study of human and rat mammary tumorigenesis. *Lab Invest* **62**:244–278, 1990.

27. Russo IH and Russo J. *In vitro* models for human breast cancer. In *Molecular Basis of Breast Cancer Prevention and Treatment.* Springer-Verlag: Heidelberg, pp. 227–280, 2004.

28. Pilat MJ, Christman JK and Brooks SC. Characterization of the estrogen receptor transfected MCF10A breast cell line 139B6. *Breast Cancer Res Treat* **37**:253–266, 1996.

29. Fernandez SV, Russo IH and Russo J. Estradiol and its metabolites 4-hydroxyestradiol and 2-hydroxyestradiol induce mutations in human breast epithelial cells. *Int J Cancer* **118**:1862–1868, 2006.

30. Sorlie T, Tibshirani R, Parker J *et al.* Repeated observation of breast tumor subtypes in independent gene expression data sets. *Proc Natl Acad Sci USA* **100**:8418–8423, 2003.

31. Christiansen JJ and Rajasekaran AK. Reassessing epithelial to mesenchymal transition as a prerequisite for carcinoma invasion and metastasis. *Cancer Res* **66**:8319–8326, 2006.

32. van de Rijn M, Perou CM, Tibshirani R *et al.* Expression of cytokeratins 17 and 5 identifies a group of breast carcinomas with poor clinical outcome. *Am J Pathol* **161**:1991–1996, 2002.

33. Russo J, Lareef MH, Tahin Q *et al.* 17β-Estradiol is carcinogenic in human breast epithelial cells. *J Steroid Biochem Mol Biol* **80**:149–162, 2002.

34. Russo J, Fernandez SV, Russo PA *et al.* 17-β-Estradiol induces transformation and tumorigenesis in human breast epithelial cells. *FASEB J* **20**:1622–1634, 2006.

35. Sorlie T, Perou CM, Tibshirani R *et al.* Gene expression patterns of breast carcinomas distinguish tumor subclasses with clinical implications. *Proc Natl Acad Sci USA* **98**:10869–10874, 2001.

36. Soule HD, Maloney TM, Wolman SR *et al.* Isolation and characterization of a spontaneously immortalized human breast epithelial cell line, MCF-10. *Cancer Res* **50**:6075–6086, 1990.

37. Bloom FE. The importance of reviewers. *Science* **283**, February 5, 1999.

38. Lawrence PA. The politics of publication. *Nature*, March 20, 2003.

39. Vogel G. Falsification charge highlights image-manipulation standards. *Science* **322**, October 17, 2008.

40. Butler D. Entire paper plagiarism caught by software. *Nature* **455**, October 9, 2008.

41. Bloom FE. The rightness of copyright. *Science* **281**, September 4, 1998.

42. Bachrach *et al.* Who should own scientific papers? *Science* **281**, September 4, 1998.

43 Russo J, Balogh GA and Russo IH. Full term pregnancy induces a specific genomic signature in the human breast. *Cancer Epidemiol Biomark Prev* **16**(1):1–16, 2008.

44. Tay LK and Russo J. 7,12-Dimethylbenz(a)anthracene (DMBA) induced DNA binding and repair synthesis in susceptible and non-susceptible mammary epithelial cells in culture. *J Natl Cancer Inst* **67**:155–161, 1981.

45. Russo J, Tay LK and Russo IH. Differentiation of the mammary gland and susceptibility to carcinogenesis. *Breast Cancer Res Treat* **2**:5–73, 1982.

46. Russo J and Russo IH. Biological and molecular bases of mammary carcinogenesis. *Lab Invest* **57**:112–137, 1987.

47. Ochieng J, Tait L and Russo J. Calcium modulation on microtubule assembly in human breast epithelial cells. *In vitro* **26**:318–324, 1990.

48. Russo IH, Koszalka MS and Russo J. Protective effect of chorionic gonadotropin on DMBA-induced mammary carcinogenesis. *Br J Cancer* 62:243–247, 1990.
49. Moral R, Wang R, Russo IH, Lamartiniere CA, Pereira J and Russo J. Effect of prenatal exposure to bisphenol A on mammary gland morphology and gene expression signature. *J Endocrinol* 196:101–112, 2007.
50. Russo IH and Russo J. The use of human chorionic gonadotropin in the prevention of breast cancer. *Women Health* 4:1–5, 2008.
51. Han HJ, Russo J, Kohwi Y and Kohwi-Shigematsu T. SATB1 reprograms gene expression to promote breast cancer metastasis. *Nature* 452:187, 2008.

Suggested Readings

Bromley DA. Staying competitive. *Science* 285:833, August 6, 1999.

Brumfiel G. Older scientists publish more papers. *Nature* 455, October 30, 2008.

Clancy CM and Eisenberg JM. Outcomes research: Measuring the end results of health care. *Science* 282:245–246, October 9, 1998.

Disappearing physician scientists. *Science* 283:791, February 5, 1999.

Fara P. Modest heroines of time and space. *Nature* 455:1177–1178, October 30, 2008.

Keller MA. Libraries in the digital future. *Science* 281:1461–1462, September 4, 1998.

Kreeger KY. Know your legal rights. *The Scientist*, March 20, 2000.

Kreeger KY. Writing Science. *The Scientist*, January 10, 2000.

Kuchinskas S. Publish thyself. *Time*, January 24, 2000.

La agonia de la poesia. *Noticias*, February 7, 1998.

Mack A. Scientific success often leads to paid public speaking engagements. *The Scientist*, p. 15, November 11, 1996.

Pinker S. Survival of the clearest. *Nature* 404:441–450, March 30, 2000.

Rensberger B. The Nature of evidence. *Science* 289:61–70, July 7, 2000.

Schiermeier Q. Self publishing editor set to retire. *Nature* 456:432, November 27, 2008.

Service RF. Pioneering physics papers under suspicion for data manipulation. *Science*, pp. 1376–1377, May 24, 2002.

Star D. Revisiting a 1930's scandal, AACR to rename a prize. *Science* 300, pp. 573–574, April 25, 2003.

Trunk P. Help! My cubemate sings show tunes. *Business 2.0*, p. 114, April 2003.

Vogelson CT. The book of knowledge. *Modern Drug Discovery*, August 2001.

Who should own scientific papers? *Science* 281:1459–1460, September 4, 1998.

Wuorio J. Keynote like a pro. *Business 2.0*, p. 52, April 2003.

Zaltman J. Getting inside-way inside-your customers head. *Business 2.0*, pp. 54–55, April 2003.

Chapter 4

The Grant Application

The Apprentice's Path to Writing a Grant Proposal

Once the scientific apprentice has found their research endeavor, in other words, the theme or the idea in which all their energies will be concentrated, its development is not crystallized in one single step. Learning about the subject is an incremental process. The scientific apprentice will quickly discover that as his or her idea unfolds, revealing even more questions and avenues for investigation, the price of conducting research becomes an important factor. Therefore it is necessary for the scientific apprentice to become well-versed in the process of procuring funds in the form of grants. The ability to generate an idea and conceptualize it to the point where it might be considered for funding is called *grantmanship*. Grantmanship is acquired by learning from experienced scientists who have been through the process, as well as cultivating the ability to persist in the face of failure (i.e. multiple rejections) and endure the arduous process of writing grants. How does the scientific apprentice face this challenge? As with most things, there is not one single, sure-fire way to ensure that a proposal will be funded. In this respect, experience is the best teacher. But in the process of learning grantmanship, looking to the ways other scientists have approached this contentious area can be helpful. In this chapter I will discuss some of the things I have learned along the way and share some collective wisdom on the topic of grantmanship, as well as how to overcome some common obstacles the scientific apprentice might face in writing a grant proposal and getting it funded.

Reasons Why the Scientific Apprentice Needs to Write Grant Applications

The scientific apprentice has identified the scientific problem, has developed a hypothesis based on his or her own investigations, and the literature support the feasibility of the idea. The problem has become ingrained in the scientific apprentice's neural network and will not at this point easily be pushed aside. The research idea, or the theme, has become a personal encounter; the scientific apprentice's relationship to their work has deepened and moved beyond superficial interest. The lingering question, also the main obstacle in the pursuit of this passion, is who will provide the funding so they can pursue this unique idea? Research grants were created to fill this need. The good news is that there are hundreds of institutions that have been created to provide grants for original research. It is in this moment in the life of the scientific apprentice that all the components discussed in the previous chapters start working together. Unless the scientific apprentice is independently wealthy, has a relative or friend willing to support their research — which even if this were case the cost of research is so high that at some point the scientific apprentice would have to seek additional funding — there is no other option other than buckling down and writing that grant application.

Beyond the practical reasons for writing grants outlined above, taking the steps necessary to write the grant application indicates an urge to fly solo. Although to say that you are alone in science is an anachronism, a creative spirit is part of our human nature. We crave to do something personal, to forge a new path not previously touched by others, and to do so on our own initiative.

Matching the Research Idea with the Objectives of the Granting Agencies

With help and direction from their mentor and peers the scientific apprentice will learn that granting institutions have different objectives in mind and particular ways in which they disburse their money. The scientific apprentice's first steps are to identify the granting institutions relevant to them, the type of funding that they provide, their overall purpose as an organization, the roster of reviewers that examine the application when available, the specific ways that funds must be used, the minimum and maximum amount of money they grant, and the application and disbursement period.

In the first step of finding the granting institution, it is important to read the institution's mission statements and the types of awards they provide. Just because many granting institutions may have mission statements which seem very similar to the scientific apprentice's interests, it may be that upon closer inspection the type of funding, or some other details, make that institution far from a good match. For example, I have found that there are dozens of agencies that are interested in providing funds for breast cancer research, but when looking for those interested in breast cancer *prevention*, the number is significantly reduced. Upon closer examination it becomes clear that from that reduced number of possible funding sources each one has a different view on what "breast cancer prevention" entails. For one agency, prevention means modifying lifestyle through diet and exercise, while for another prevention means early diagnosis, or screening. Some funding agencies also see prevention in terms of combating potentially noxious environmental factors. These variations within the same field of study narrow the pool of possible funding sources considerably. The scientific apprentice quickly learns that the list of granting agencies gets shorter and shorter, but the search for the granting agencies must nevertheless continue, however challenging. The reality is that these agencies exist and their purpose is to provide funds. The first struggle is identifying them.

Once the scientific apprentice has identified the appropriate agency, the next question that needs to be asked is how many applications that agency receives and funds per year? The best estimation of grant success is to find out how many applications have been received by the agency and how many are actually funded. A rate of 25% funded applications suggests that applicants have a strong chance of getting funded, but if the funding rate drops to 5% or less it means these awards are highly competitive, therefore decreasing the odds that a scientific apprentice's project will be selected.

Most granting institutions such as the National Institute of Health have a fixed board of reviewers that are rotated periodically and their names are available to the public. Other institutions, such as the Department of Defense, do not have a fixed roster of reviewers, rather they are selected experts chosen to fit the needs of specific areas. Knowing the names of possible reviewers of your grant application is helpful, to a certain extent, and could help in getting an idea how the grant will be reviewed. However, the process remains unpredictable so it is important not to get hung up on assumptions about how any particular reviewer will react.

Lastly, the scientific apprentice must learn how long the agency takes from the time that the application is received to the time of announcement of the award. Some institutions take only a few months while others can take up to nine months before giving a final answer.

When possible, it is important to meet or talk to the person in charge of the granting agency, especially the directors who handle the program. These people are extremely helpful in providing accurate information and can answer most of the questions outlined above. As in all situations it is possible that you will come across some less than helpful individuals, but for the most part I have found that people have a genuine interest in research and a belief that the best way to solve the problem of breast cancer (or any other scientific question) is through the grant mechanism. Unfortunately, there are also some program directors who maintain their position by encouraging new scientists to apply out of a sense of self preservation, not out of genuine desire to help the applicant.

How to Write a Grant Application

While each granting agency has a defined format for presenting the application, here I will concentrate on the Research Grant (RO1) application. The formats vary but the basic principles are the same. The RO1 application is a research project initiated by the investigator and is most highly regarded when granted by the National Institute of Health. This type of grant proposal, to a certain extent, defines the research scientist because this type of grant application does not have boundaries, only those imposed by the researcher him or herself.

The grant application has four basic parts: *Specific Aims, Background and Rationale, Preliminary Results* and *Methods and Procedures*. In this type of application, each of these parts are intertwined and all of them need to be strong in order to be considered fundable. Although the specific aims are the first part of the application, the writing begins with the Preliminary Results. This part supports the aims and rationale by explaining why one specific approach will work over another, which must then be substantiated with a solid Methods and Procedures section that contains the type of experiments proposed. After the preliminary results are written, the next step is to develop the Specific Aims, followed by the Methods and Procedures.

The Methods and Procedures section is not simply a list of recipes but a disquisition on how the experiments will be designed and what methodology will be used. When writing this portion of the application, useful questions to consider are: Is this methodology new? Is the investigator proficient in the use of the techniques proposed? What do we expect to gain with the approach proposed? What if the experiment fails? What are some alternative strategies to follow? Is the statistical portion strong enough so that the data will be a step forward and not a fishing expedition? All of these questions must be kept in mind when writing a grant application.

How I Wrote My Grant Application

Perhaps the best way to understand the architecture of an RO1 grant is to look at an example of a successfully funded project. I will take the reader through one of my own National Cancer Institute funded RO1 applications. Although several years old now it should still prove to be a useful learning tool.

Specific Aims

In this application I have two specific aims:

Specific Aim 1: To characterize the specific gene expression profile of women at "low" and "high" risk of developing breast cancer because of reproductive history.

Specific Aim 2: To determine if gene clusters differentially expressed in women at risk of developing breast cancer because of reproductive history (identified in Specific Aim 1) are differentially expressed in the breast tissue of postmenopausal women with breast cancer compared to postmenopausal women who do not have breast cancer.

For those in the field of breast cancer and who know the preventive effect of early first full term pregnancy, the specific aims as stated may mean something. However, in a grant application the scientific apprentice must not assume absolutely anything. The scientific apprentice must guide the reviewers step by step, like a well-crafted novel. Why write in this way? Reviewers, like you, have weaknesses, and even though they may know a substantial amount about the subject, you are presenting them with a new paradigm that they need to place in the framework of their own knowledge. Therefore the application must be engaging and must convey its value.

Once the scientific apprentice has captured the reviewer's attention, the next step is to captivate their imagination. The scientific apprentice's idea must be eloquently presented and fend for itself among the hundreds of applications. Therefore the Specific Aims must begin as follows:

Breast cancer is the most common neoplastic disease in women and accounts for up to one third of all new cancer cases in North American women. Breast cancer mortality has remained almost unchanged for the past five decades, and is second to lung cancer as a cancer-related cause of death.[1]

This first paragraph has stated the problem, and if the reviewer is not familiar with the data it will clearly illustrate that breast cancer is not a trivial problem and affects a wide range of people, maybe even someone near and dear to the reviewer. In the event that the reviewers are already well-versed in the facts laid out in this section, the scientific apprentice still wants to keep their attention. Use the second paragraph to provide an additional reason why the study of breast cancer is important by adding:

Our failure to eradicate this disease is largely due to our inability to identify a specific etiologic agent, determine the precise time of initiation, and resolve the molecular mechanisms responsible for cancer initiation and progression.

I would advise not to dwell too much on the epidemiology at this point of the application. A few examples are enough before moving onto the next point. Start by providing a sense of hope and that science has not been completely indifferent to the problem and by framing the problem toward the area that the scientific apprentice is interested, in this case the endocrinological aspect of the disease.

> Despite numerous uncertainties surrounding the origin of cancer, there is substantial evidence that breast cancer risk relates to endocrinological and reproductive factors. Breast cancer development is strongly dependent on the ovary and on endocrine conditions modulated by ovarian function, such as early menarche, late menopause, and parity.[1-5]

In the next paragraph, the scientific apprentice must define where the current research stands and what aspect of the specific area of research they want to emphasize. This paragraph is important because the scientific apprentice needs to demonstrate that they know what they are saying in less than 200 words. The scientific apprentice has generated knowledge based on findings already published in peer review journals; he or she must therefore prove their thorough understanding of the existing work before expressing the specific aims of their proposal, by pointing out that,

> Epidemiological findings indicate that the incidence of breast cancer is higher in nulliparous women, while a lifetime risk decrease has been observed in those parous women whose first pregnancy was completed before age 24.[2-7] The protection conferred by pregnancy has been in great part explained by our studies of 7,12-dimethyl benz(a)anthracene (DMBA)–induced carcinogenesis. In this model, full term pregnancy practically abolishes mammary cancer development,[6,8-15] a phenomenon mediated by modifications in the pattern of lobular development and differentiation, cell proliferation, steroid hormone receptor content, and gene expression. Changes observed in rodents are similar to those induced by pregnancy in humans.[16-22]

The last sentence, "Changes observed in rodents are similar to those induced by pregnancy in humans," establishes the link between the experimental facts and what is observed in humans. The scientific apprentice must continue elaborating on the human data:

> In women an early first full term pregnancy (FFTP) confers protection, however, delay of childbearing beyond age 35 progressively increases the risk of cancer development. When the FFTP occurs after 35 years of age, a woman's risk of developing breast cancer is greater than that of nulliparous women.[2-5]

This is the point where the scientific apprentice must establish authority in the field by presenting their own hypothesis and a demonstration that this hypothesis might be the answer to the problem:

> We have called the period encompassed between menarche and the FFTP the **"susceptibility window"**; we postulate that the larger this window the greater the risk of the breast being damaged by radiation, environmental carcinogens, and/or hormonal imbalances. Genomic damage caused by any one or several given agents might vary as well, depending upon the genetic predisposition of the patient. Thus, late pregnancy might stimulate the growth of foci of transformed cells already present in the breast at the time of conception.

In the previous paragraph, the scientific apprentice has stated the main ideas behind the work that will be presented in greater detail than in the preliminary part of the application, however, the scientific apprentice must not leave any doubt about whether all other possible explanations to the problem have been contemplated:

> Another alternative is that the breast of women becomes refractory to undergoing complete differentiation with aging.[16,17,22-28] This possibility is supported by our observations that the breast of parous women with cancer has an architectural pattern of development similar to that of nulliparous women, suggesting that the breast of the former might have

failed to differentiate under the influence of the hormones of pregnancy.[16,17,27] Genetic influences that are responsible for at least 5% of all the breast cancer cases also seem to influence the pattern of breast development and differentiation.[17] The fact that early FFTP confers life-long protection from breast cancer, whereas a greater risk is associated with nulliparity, clearly define a "low risk" and a "high risk" population.

In the sentence, "The fact that early FFTP confers life-long protection from breast cancer, whereas a greater risk is associated with nulliparity, clearly define a 'low risk' and a 'high risk' population," defines the population of the study. From here we move to the methodology used in the study, highlighting that the scientific apprentice is using a state-of-the-art technology:

> The advances of the human genome project and the availability of new tools for genomic analysis, such as cDNA array, tissue array, laser capture microdissection (LCM), and bioinformatic techniques allow us to determine whether there are clusters of genes that are differentially expressed in populations that differ in their breast cancer risk. Furthermore, those clusters of genes whose expression may be affected by early pregnancy and that can be proven to be functionally relevant in protecting the breast from cancer development could serve as markers for evaluating cancer risk in large populations.

Although the idea in the paragraph is complete, it lacks power. By adding,

> This concept has been proven to be correct in the rat model in which a cluster of genes remain activated after the process of involution postpregnancy takes place, conferring a **"special genomic signature"** to the gland that is responsible for its refractoriness to chemical carcinogenesis.

The scientific apprentice is giving in this paragraph two important messages: that the hypothesis' applicability to humans is not an afterthought, but the consequence of carefully designed experimental animal models, and that the scientific apprentice has established a reputation in the field of prevention.

This is the moment to make the transition from a discussion of the animal models, to human, concluding the idea by explaining why this moment is the right one to test the hypothesis:

> It is on this ground that we propose to test the hypothesis that early pregnancy imprints in the human breast permanent genomic changes or a "signature" that reduce the susceptibility of the organ to undergo neoplastic transformation. If this hypothesis is true, it is expected that the cluster of genes associated with early FFTP would be absent or modified in the breast of high-risk women, i.e., nulliparous women.

The scientific apprentice must then define which specific group of women the study will focus on:

> For this purpose we have selected for our study postmenopausal women 50 to 65 years old. These women have the dual advantage of being free of cyclic hormonal variations that occur in premenopausal women, and have permanently imprinted genomic changes or signatures in the mammary epithelium related to an early reproductive event.

It is only at this point of the first part of the Specific Aims section of the application that the reviewers are ready to hear what the aims of the study are:

> For testing this hypothesis we propose the following two specific aims:
>
> **Specific Aim 1:** To characterize the specific gene expression profile of women at "low" and "high" risk of developing breast cancer because of reproductive history.

The scientific apprentice clearly indicated that there are two specific aims. After the first aim is defined, the scientific apprentice summarizes how it will be carried out by saying:

> Gene expression in normal breast tissue obtained from postmenopausal women with a history of one or more early full term

pregnancies will be compared to gene expression in normal breast tissue from postmenopausal women who are nulliparous. For this purpose we will perform a large-scale analysis of gene expression using available human cDNA libraries for determining the expression pattern of known and new (ESTs) genes. Differential expression of gene clusters in the "low" and "high" risk groups will be determined using discriminant analysis to allow for adjustment for age and race. We will confirm the expression of the identified gene clusters utilizing real-time quantitative PCR and/or *in situ* hybridization or immunocytochemistry, when the antibody for the coded gene is available on constructed tissue arrays with the paraffin embedded tissue of the samples included in this study.

When writing the Specific Aims, it is important to maintain parallel sentence structure and uniformity in the text by describing the second specific aims in the same fashion as the first:

Specific Aim 2: To determine if gene clusters differentially expressed in women at risk of developing breast cancer because of reproductive history (identified in Specific Aim 1) are differentially expressed in the breast tissue of postmenopausal women with breast cancer compared to postmenopausal women who do not have breast cancer. For this purpose we will perform a case-control study to compare gene expression profiles in the normal breast tissue of postmenopausal women with invasive breast cancer vs. postmenopausal women with benign breast disease.

Thus far the scientific apprentice will have described in about 1000 words the grant proposal that he or she is planning to present in greater detail in the following 24 pages or less of the application. The importance of the Specific Aims cannot be overemphasized. Specific Aims are the first thing the reviewer will read and if the whole proposal is not clearly stated here, the reviewers will need to dig into the following pages in order to make sense of the general motif of the application. The Specific Aims must not only spark the reviewer's interest, but also make them eager to continue reading the application.

Background and Significance

The function of the Background and Significance section of the application places the applicant's idea into the context of available knowledge. This section serves as the frame of the entire project. While there may be paragraphs or sentences that rehash ideas already expressed in the Specific Aims, here these ideas and references are elaborated and explained in greater detail. The purpose of the Background and Significance section is to establish a dialogue with the reviewers by explaining how the idea originated based on the problem it intends to solve. For example:

> The incidence of breast cancer has gradually increased in the United States and in most Western societies over the last few decades.[1] Although the reasons for this increase are not certain, epidemiological, clinical, and experimental data indicate that the risk of developing breast cancer is strongly dependent on the ovary and on endocrine conditions modulated by ovarian function, such as early menarche, late menopause, and parity.[1-5] It is paradoxical that age is another important risk factor, because breast cancer, which is practically non-existent before age 24, exhibits maximal incidence during the postmenopausal years, when the ovaries are not longer functional.[1-5] The majority of breast cancer patients are women in the sixth and seventh decades of life, and the mortality for breast cancer also continues to rise after menopause. The age-specific incidence, i.e., the number of cases per year per 100,000 women in each age group, climbs rapidly after the age of 30 years, reaching a peak of maximal incidence of 500 cases per 100,000 women in the 60–70 years old group.[1] The fact that women who gave birth to a child when they were younger than 24 years of age exhibit a decrease in their lifetime risk of developing breast cancer, and that additional pregnancies increase the protection,[24] adds complexity to this paradox. A plausible explanation for the lifetime protective effect of an event occurring so early in life is provided by the biological behavior of breast cancer and by comparative studies with experimental animal models. Epidemiological observations indicate that a higher breast cancer incidence occurs in women who had been irradiated, but only in those in whom exposure occurred at a young age, particularly before 19 years of age, but not in those that were irradiated at older ages or after pregnancy.[29]

The paragraph above explains the epidemiology of the disease and the basis for studying the problem. It is important however that the scientific apprentice alert the reviewer to the applicant's work. The scientific apprentice must not forget that one of the scientific apprentice's tasks is to demonstrate how they are part of the solution to the problem, and prove to the reviewers they have made contributions which have advanced the field. To do this the scientific apprentice must indicate how their previous work supports the present application. It is important to emphasize here that the RO1 application requires having preliminary data and it is advisable to present the work carried out by the scientific apprentice in an objective manner:

> In rodents, maximal incidence of 7,12-dimethylbenz(a)anthracene (DMBA) induced mammary cancer occurs when the carcinogen is administered to young virgin cycling rats, but the same carcinogen fails to induce tumors when given to rats after a full term pregnancy.[6,23,28] The high susceptibility of the young virgin rat mammary gland to develop malignancies is the result of the interaction of the carcinogen with rapidly dividing cells present in terminal end buds (TEBs), undifferentiated structures that represent the most active growth centers of the mammary parenchyma. Cancer initiation in this model is the result of a combination of a high rate of carcinogen binding to the epithelial DNA, fixation of transformation, formation of polar metabolites, and deficient DNA repair,[11,12,30–34] among other factors.

This citation can be reinforced by mentioning scientific apprentice's other work with the objective of establishing authority in the field. If the data came from another angle, such as from observations in humans rather than animal, this is even better. For example:

> Although no specific etiologic agent for breast cancer has been identified, there are close similarities between the pathogenesis of this disease in women and that induced in rodents by chemical carcinogens. Ductal carcinoma, the most common breast malignancy, originates in lobules type 1 (Lob 1), also called terminal ductal lobular unit (TDLU), an undifferentiated structure that is considered to be equivalent to the TEB, the site of origin of ductal carcinomas in rodents.[8,9,14,16,17,25]

Because this grant application seeks to reinforce the importance of breast cancer's site of origin, sustaining the hypothesis that lobule type 1's genomic signature is significant, the *in vitro* study points out to the reviewer how the experiment supports the concept:

> Further, under *in vitro* conditions the same chemical carcinogens that induce mammary cancer in experimental animals[13,14] can transform human breast epithelial cells. These observations suggest that if the human breast is exposed to a carcinogenic insult, the Lob 1 or terminal ductal lobular unit (TDLU) would be the structure affected, and thus the site of initiation of a malignancy.[8]

The following paragraph points out how the genomic signature could be affected by other factors:

> Therefore, it is possible to postulate that genomic damage caused by radiation, environmental carcinogens, hormonal imbalances, and/or other still unidentified factors, either alone or in combination with genetic predisposition, might cause breast cancer in women. For cancer to develop, however, this multifactorial combination must occur during the window of high susceptibility that is encompassed between menarche and the first full term pregnancy (FFTP), even though the damaged cells would be clinically detectable as a neoplasm only after several years of progressing along the various stages of transformation.[13–15] An understanding of how these multifactorial influences lead to genomic damage and cancer initiation requires identifying within the breast the foci of high susceptibility to be transformed.

The applicant should continue focusing on the difficulties that have caused delays in the generation of this knowledge, and can enlighten the reviewer in this respect by including:

> This objective is marred by innumerable difficulties, such as the large size of the organ and complex parenchymal structure. The fact that the mammary gland is one of the few organs that is not fully developed at birth

represents the first level of difficulty.[21,22] No other organ presents such dramatic changes in size, shape and function, as does the breast during growth, puberty, pregnancy, lactation and postmenopausal regression.[21,34–37] Breast development starts as early as the stage of nipple epithelium during embryonic life, and continues after birth in parallel with body growth. A spurt of growth with lobule formation occurs at puberty, but the completion of breast development and differentiation occurs only at the end of a full term pregnancy.[21,37] Thus, as long as a woman does not become pregnant the predominant structure in her breast is Lob 1. With pregnancy the mammary parenchyma reaches the final stage of secretory lobule type 4 (Lob 4) that forms by the end of the reproductive process and remains present during lactation. Despite undergoing regression after weaning, the breast of parous women retains until the third and fourth decade of life differentiated lobule type 3 (Lob 3) as the predominant structure. Lob 3 has a lower rate of cell proliferation and steroid hormone receptor content than Lob 1, and expresses several genes that are related to the differentiation process,[18–20] such as a new serpin gene,[19] mammary derived growth factor inhibitor,[20] and genes controlling programmed cell death and DNA repair.

The scientific apprentice must continue discussing how the interpretation of the protective effect of pregnancy in humans can also be explained by the use of experimental animals. This idea can be further expanded:

> The protection conferred by pregnancy has been, in great part, explained by our studies of DMBA-induced carcinogenesis. In this model, an almost complete abolition of the oncogenic response to DMBA results from the induction of differentiation of the mammary gland by either full term pregnancy or treatment of virgin rats with human chorionic gonadotropin (hCG), one of the main hormones produced during pregnancy.[15,23]

The above paragraph highlights the pioneering concept that differentiation is the driving force behind the protective effect of pregnancy. Whereas we have chosen differentiation as a concept, this can also apply to other mechanisms such as chromatin remodeling or epigenetic silencing. Whatever the topic, it is worth defining it in terms of its application. Because the word *differentiation* could have several different meanings, the following puts the term in the correct context and defines it in relation to mammary development:

> We define differentiation as the coordinated and sequential series of events induced in the breast by the hormonal milieu of pregnancy, or pregnancy-like conditions, that culminate in the activation of genes controlling ductal and lobular development inducing a unique genomic signature.

In the next section of Background and Significance the scientific apprentice must provide a summary of findings to further illustrate to the reviewers that the rationale of the project is based on data collected from various sources. The following paragraph explains how genomic arrays will be the main technique used in project and that the information obtained from this methodology will help to strengthen the hypothesis.

> We have compared gene expression in breast epithelial cells obtained from Lob 1 and Lob 3 of nulliparous and parous women, respectively. Analysis, utilizing a nylon filter membrane that contains 1,176 genes, revealed that in the more differentiated Lob 3 genes that have never been considered relevant in the normal breast are upregulated, thus providing a new insight in their role in the differentiation pattern of this organ (See Preliminary Studies). It is of interest to note that known genes that are overexpressed in breast cancer, such as HER-2/neu[38] and mucin[39,40] are not expressed in normal lobular structures.

The applicant should then proceed to establish the relationship of the findings described at the molecular level to previous studies showing the response *in vitro* of the lobular structures type 1 and 2 to the carcinogenic challenge of these two different stages of cell differentiation:

> In addition, cells derived from the differentiated Lob 3 are resistant to growth *in vitro* and do not express transformation phenotypes upon carcinogen treatment, as cells from Lob 1 do.[13,14,21,22] The protection conferred by an early pregnancy[1,3,14,42,43] is mediated by the induction of differentiation

of the breast by the reproductive event, as it has been demonstrated in the rodent experimental model.[15,23,25,28,30,31,41]

Next the applicant wants to reiterate the fact that there have been exceptions reported regarding the protective effect of pregnancy and that those exceptions are not contradictory to the main hypothesis:

The fact that a certain percentage of parous women develop breast cancer[24] could be explained as a defective response of the breast to the hormones of pregnancy. This postulate is supported by differences in the architectural pattern of breast tissues from parous cancer patients, which appears similar to that of nulliparous women, having the Lob 1 as the prevalent structure.[16,17,27] The possibility that these breast tissues exhibit a defective response to the differentiating influence of the hormones of pregnancy warrants further investigation. The pattern of breast development and differentiation is also influenced by inheritance.[17,22]

It is important to keep in mind that at this point the applicant is maintaining a dialogue and a slightly speculative tone could be useful:

Therefore, although there is no explanation as of yet for the higher breast cancer risk exhibited by nulliparous and late parous women, the fact that experimentally induced rat mammary carcinomas develop only when the carcinogen interacts with the undifferentiated and highly proliferating mammary epithelium of young nulliparous rats,[28,30,32,41] suggests that the breast of these "high breast cancer risk" women might exhibit some of the undifferentiated cell characteristics that have been shown to be essential for the initiation of cancer in rodents. The higher proliferative activity of the nulliparous woman Lob 1, in association with the higher breast cancer incidence in this group of women, suggest that these lobules are biologically different from those of early parous women.[13,14,18,21,22,25]

At the end of this paragraph the applicant might continue in the same speculative voice, only now with the inclusion of a description of which groups of women will be investigated and an elaboration of what was written in the Specific Aims section. This last part also seeks to outline the expected findings of the study:

> We seek to confirm this postulate by determining whether in post-menopausal women with a history of early FFTP the breast epithelial and stromal cells exhibit a "genomic signature" or gene expression different from those of the breast tissue from nulliparous women. Based on our preliminary studies, we expect that genes that control cell proliferation will have lower expression, whereas genes controlling programmed cell death, DNA repair and cell differentiation will have higher expression in Lob 1 of early parous postmenopausal women, whom we have defined as the "low breast cancer risk" group. The recent data obtained in the rat model in which cluster of genes remain activated in the involuted gland at postpregnancy, conferring a "special genomic signature" provide a powerful rationale for characterizing the genomic signature of breast tissue of postmenopausal women.

Approaching the end of this section's three-page limit is the applicant's final opportunity to underscore the importance of methodology that has already been mentioned, but may need to be better rationalized. At this point the applicant must also present an argument that defends the method as an ideal way to demonstrate the main hypothesis. At the time this example was written (in year 2000) many of the techniques being justified were just emerging. The apprentice of science must emphasize the technological advances relevant to his or her time when writing grant applications. The narrative that follows from the final part of the Background and Significance section demonstrates how developing technology is relevant to the success of the project:

> The emerging new technology of DNA microarrays, developed on the basis of recent technological achievements and the sequence information generated by the Human Genome Project, allows one to analyze simultaneously the expression of thousands of genes in a given cell population.[44-47] Equivalent amounts (~5 ng) of PCR-amplified DNA sequences (0.5–2.0 kb) of specific cDNA clones are immobilized at low density (10–20 cDNAs/cm^2) on nylon-membrane filters (macroarrays),

or at high density (1,000–6,000 cDNAs/cm^2 or $\geq$ 250,000 groups of ~10^7 oligonucleotides/cm^2) through mechanical microspotting onto poly-L-lysine-coated glass microslides, or *de novo* synthesized *in silico* by photolithographic methods.[48,49] Complex cDNA probes are derived *in vitro* by reverse transcription of purified total or, preferentially, poly (A)+RNAs, and an incorporated radioactive (^{32}P or ^{33}P) or fluorescent (Cy3 and Cy5) label, which are then hybridized directly to the immobilized DNAs. The detection of the hybridized DNA probe is usually carried out by means of a phosphorimage scanner or direct autoradiography in the case of microarrays hybridized with a radioactive probe, or a laser confocal scanner for fluorescence-labeled probes.

The narrative of the new methodology should also emphasize how the data will be analyzed.

A problem that most investigators usually have to deal with is the reproducibility and reliability of extracting gene expression array data within a linear range, as well as the difficulty in handling and interpreting the large and complex data sets generated. While algorithms for processing the obtained information are currently available at the level of comparing gene-to-gene and array-to-array, only a limited number of statistical packages have been developed to handle the analysis and interpretation of the array data to its full complexity. Ideally, an investigator would like to have the ability to analyze multiple data sets simultaneously, and to be able to extract pattern information.

The applicant wants to indicate that they have the statistical analysis of the genes covered by explaining:

Such analysis requires powerful, multivariate statistical approaches, which seek to identify common gene patterns that characterize a given group of cells undergoing discrete phenotypic changes. Employing such algorithms should significantly increase our ability to maximally extract information from gene-array data sets, and should allow the prediction of statistically probable gene expression patterns, characteristic of cells traversing the continuum between normal and malignant, or responding to specific growth conditions. A number of statistical approaches for clustering gene expression profiles have been recently applied by various investigators to analyze array data from 8,000–10,000 gene arrays in diverse studies, including cell cycle progression and the induction of stress in yeast,[50,51] serum stimulation of primary human fibroblasts,[52] as well as the molecular classification of tumor tissues and cell lines.[53–61]

The applicant also wants to emphasize that the technique in itself, although important, is sustained by solid analytical power. The following paragraph describes how the data will be analyzed and interpreted:

> Multivariate statistical techniques, such as Cluster Analysis and Self Organizing Maps (SOM), were successfully used in these cases. Other multivariate statistical techniques with powerful pattern recognition capabilities that could be used for reducing the dimensionality of the gene array data down to a finite number of gene clusters, have been successfully implemented at Fox Chase Cancer Center (FCCC) for extracting well-defined, statistically significant patterns from a variety of NMR spectra and images, as well as gene expression array data (See Preliminary Studies), including Principal Component Analysis (PCA), and Bayesian analysis with AutoClass.[62] The use of microarray techniques will also allow us to identify anonymous ESTs that cluster with genes of known function, thus allowing their subsequent analysis in a more detailed and informed manner based on their potential function.

The applicant also wants to assure the reviewers that he or she realizes, as shown in our example, that the list of genes identified and statistical analysis of them will not suffice unless a second method of verification, or validation, of the data are completed.

> The validation of known and unknown genes by QRT-PCR, *in situ* hybridization or immunocytochemistry utilizing tissue arrays of the examined samples will also provide another functional dimension to these studies. It is important to indicate that, despite the exploratory, model-independent

nature of a genome-wide expression profiling, as has been emphasized by Brown and Botstein,[61] the expression profiles sought by this application are in essence molecular phenotypes that should shed light on the functional differences between the breast tissue of women with a defined risk to develop breast cancer. This strategy is the only one presently available that could provide information on existing differences that are not obvious at the morphologic level.

The conclusion of the Background and Significance section should show confidence that, having explained the basis and rationale for the proposed hypothesis, the applicant has a clear vision for how their research will affect their field:

> The knowledge gained through these studies will result in a better understanding of the molecular basis underlying morphological differences that are associated with the physiological process of pregnancy, and perhaps responsible for the differences in cancer incidence between early parous women and nulliparous women.

Lastly, the applicant wants to further stress the overarching aim of the application:

> In addition, the characterization of the signatures of gene expression in the "low risk" population will serve as intermediate end points useful for evaluating breast development, the response of the breast to chemopreventive agents or hormones, and/or susceptibility to malignant transformation.

It is important to close this section by keeping the gestalt of the project at the forefront of the reviewer's mind. The reviewer must understand the importance of the problem, that the aims are sound and that the rational behind the project is strong. They must also feel that the application is innovative and significant, and if accomplished will be of great impact. If this is achieved the reviewer will already have a score in mind in the "outstanding" range. The rest of the application must only confirm what the reviewer has already concluded thus far.

Preliminary Studies

In the Preliminary Studies section of the grant application the applicant provides the data that will sustain the hypothesis. It could be that the specific aims and the background and rationale are appealing to the reviewers, but it is the preliminary data that will lend credibility to the hypothesis and will demonstrate that the applicant is ready to test it. For the sake of completeness, I would like to share with the reader how in this application model the editing and layout of the preliminary results that support the hypothesis were developed in the grant. It is the applicant's responsibility to make it easy to connect the relevant points of each publication that the hypothesis is based on by presenting the data in a way that shows the applicant has digested and understood the studies.

I suggest beginning the Preliminary Results section with an introduction that prepares the reviewers for the scope of the data that the applicant will present:

> Our studies have been designed with the purpose of understanding the developmental and molecular mechanisms through which pregnancy and hormonally induced differentiation affect the mammary gland, and its relationship with the pathogenesis of the disease, to determine whether genomic differences exist between the nulliparous and parous breast, and to develop the technical approaches to be applied in this application.

In our example, it is crucial that the concept of differentiation of the breast as the cornerstone of the genomic signature of protection induced by parity is understood, therefore a summary of the morphological studies that are the basis for the development of the hypothesis is necessary.

> i. **Architecture of the human breast.** The breast progressively develops from infancy to puberty under the main stimuli of pituitary and ovarian hormones. The least differentiated structure identified in the breast of postpubertal nulliparous women is the Lob 1, or TDLU, which is composed of clusters of 6 to 11 ductules per lobule. Lob 1 progress to Lob 2, which are composed of more numerous ductules per lobule and exhibit a

more complex morphology. A fully differentiated condition is reached by the end of a full term pregnancy, under the stimulus of new endocrine organs, the placenta and the developing fetus. These new hormonal influences induce a profuse branching of the mammary parenchyma leading to the formation of secretory lobular structures. During the first and second trimesters of pregnancy Lob 1 and Lob 2 rapidly progress to Lob 3, which are composed of more numerous and smaller alveoli per lobule. During the last trimester of pregnancy, active milk secretion supervenes, the alveoli become distended, and the lobules acquire the characteristic of Lob 4 which is present during the lactational period. After weaning, all the secretory units of the breast regress, reverting to Lob 3 and Lob 2.[21,22,25]

Incorporating figures or diagrams that make it easier to illustrate the idea is fundamental. As always, the applicant must help the reviewers understand the basis of the hypothesis and adequate graphic materials are helpful to this end. It could be new graphic material generated by the applicant, or better yet, figures obtained from the original publications that indicate the source. In some applications, inclusion of an appendix section is permitted if additional information is needed that supports the contention of the hypothesis to be tested. In the example provided the applicant also introduces, in addition to the morphological description of the breast, useful information about proliferate activity and its relation to the presence of hormone receptors as follows:

ii. Cell proliferation and steroid receptor content in the normal breast. The initial classification of lobules into four categories, primarily based on morphological characteristics of these structures[21,22,25] was complemented by analysis of the rate of cell proliferation, a cellular function essential for normal growth.[18,63,64] Normal growth requires a net increase of cycling cells over two other cell populations, resting cells (arrested in G_0), and dying cells (cells lost through programmed cell death or apoptosis). Proliferating cells express the nuclear protein Ki67, which are detected with a monoclonal antibody against it. We determined that the proliferative activity of the mammary epithelium varies as a function of the degree of lobular differentiation. The percentage of Ki67 positive cells (Ki67 or proliferation index) decreases progressively as the lobules mature from Lob 1 to Lob 2, and these to Lob 3 and Lob 4.[18,63,64] These differences were not abrogated when the proliferation index was corrected for the phase of the menstrual cycle.[63,64] We concluded that parity, in addition to exerting an important influence in the lobular composition of the breast, profoundly influences the proliferative activity of the breast. Estrogens and progesterone are known to promote proliferation and differentiation in the normal breast. Both steroids act intracellularly through nuclear receptors, which become activated by the binding of their respective ligands. This is the most widely accepted model of action of estrogens for inducing cell proliferation and regulating gene expression.[65–75] We have utilized monoclonal antibodies that specifically recognize estrogen receptor alpha (ERa) since the discovery of the ER-beta,[18,63,64] and progesterone receptor (PgR) in normal breast tissue (see Appendix, Exhibit B). Our studies revealed that the proliferative activity and percentage of cells positive for both ER- alpha and PgR are highest in Lob 1, and they progressively decrease in an inverse relationship to the degree of lobular differentiation, providing a mechanistic explanation for the higher susceptibility of these structures to be transformed by chemical carcinogens *in vitro*.[13,14]

Above, the applicant has selected a specific set of data that the reviewers will use to build the basic knowledge he or she needs to have at hand to evaluate the work proposed. The next paragraph accomplishes this by describing the influence of age and parity on the breast. The study in question will use postmenopausal breast tissue, so it is relevant that the reviewers understand what the breast looks like after the process of pregnancy and lactation and what it looks like at menopause, after the process of involution.

iii. **Influence of age and parity on breast development and cancer susceptibility.** Our studies of chemically induced carcinogenesis in an experimental animal model have shown that the initiation of the neoplastic process is inversely related to the degree of differentiation of the mammary gland, which in turn is a function of age and reproductive history.[6,8,23,76] These observations led us to hypothesize that the protective effect of early first full term pregnancy in women is the result of the differentiation of the breast.[13,14] To test this hypothesis we compared the architecture of breast tissues obtained from reduction mammoplasties of nulliparous and parous women. The breast of nulliparous women contained almost exclusively Lob 1, and their number remained nearly constant throughout the lifespan of the individual. The breast of early parous women contained predominantly the more differentiated Lob 3, whereas Lob 1 were in a very low percentage until the fourth decade of life, when they started to increase, reaching the same level observed in nulliparous women after menopause. The breast of nulliparous women never reached the degree of differentiation found in women who completed an early pregnancy. Even though during the postmenopausal years the preponderant structure is Lob 1 in the breast of both parous and nulliparous women, only nulliparous women are at high risk of developing breast cancer, whereas parous women remain protected.[21] Since ductal breast cancer originates in Lob 1 (TDLU),[8,9] the epidemiological observation that nulliparous women exhibit a higher incidence of breast cancer than parous women[3,4] indicates that Lob 1 in these two groups of women might be biologically different, or exhibit different susceptibility to carcinogenesis.[14,21,22] Even though Lob 1 is the hallmark of the postmenopausal breast, we postulate that the degree of differentiation acquired through early pregnancy has caused a "genomic signature" that differentiates Lob 1 from the early parous women from that of the nulliparous women. The identification of such "molecular phenotypes" will shed light on the functional differences between cells and tissues that may not be obvious at the morphological level. Furthermore, we expect that a broad mapping of gene expression of early parous women free of cancer in comparison with that of nulliparous or women with cancer will allow us to identify the differences postulated here. Genomic mapping will significantly add to our observations of lower proliferative activity in the Lob 1 of the parous women's breast, and higher in the Lob 1 of the nulliparous women's breast.[63,64] In addition to the differences in proliferative activity, the three types of lobules exhibit variations in their *in vitro* growth characteristics. Lob 1 and Lob 2 which grow faster have a higher DNA labeling index, and a shorter doubling time than Lob 3.[43] They also exhibit different susceptibility to carcinogenesis.[13,14] Cells obtained from Lob 1 and Lob 2 express *in vitro* phenotypes indicative of neoplastic transformation when treated with chemical carcinogens, whereas cells obtained from Lob 3 do not manifest those changes.[13,14]

The applicant must continue by summarizing the previous concepts and give additional evidence by introducing relevant experimental data:

The confirmation in an experimental system that pregnancy and human chorionic gonadotropin (hCG) treatment confers long lasting protection, clearly indicates that the differentiation induced by these processes is a permanent modification of the biological characteristics of the mammary gland, in spite of the regression of differentiated structures to seemingly more primitive conditions.[23,76-78] Collectively, our data establish a baseline for understanding the evolution of glandular development, and how age and parity influence it. This knowledge is of utmost importance for understanding the role of differentiation in the protection of the mammary gland against carcinogenesis.[25,76] In addition, these data establish well-defined endpoints for studying the response of the mammary gland to hormonal or chemopreventive agents, which could be utilized in modulating the susceptibility of the breast to carcinogenesis.

The following data, which highlight the importance of the parenchyma-stroma relationship, introduces new background information generated in the laboratory. This is relevant because the application proposes studying the epithelia (or parenchyma) and the stroma of the human breast. It is therefore crucial that the applicant establish the basis for the experimental approach and the morphological substratum of the need to do laser capture microdissection, as mentioned in the methods, to separate the cells from the parenchyma and those from the stroma.

iv. **Parenchyma-stroma relationship.** Breast development occurs through a process of ductal elongation, branching and sprouting of ductules or alveoli, a process that requires extensive cell proliferation and penetration of the ductal epithelium into the stroma.[21] Both the intralobular and the interlobular stroma are affected simultaneously during development, pregnancy, lactation, and involution.[21,79,80] These processes occur, in turn, in a synchronous manner in response to specific hormonal and growth factor stimuli.[37] Two major mechanisms are considered to be involved in the interaction of the stroma and epithelial cells, the production of soluble growth factors, a modification of the composition of the extracellular matrix. This interaction seems to be bidirectional such that epithelial cells are also capable of influencing stromal cell behavior and governing gene expression.[79,80] The study of the stroma-parenchyma ratio in 14 breasts of pubertal, postpubertal, parous, and pregnant women, showed that the relationship between parenchyma and stroma is a dynamic process.

The applicant concluded this section by providing a further explanation why the stroma and the parenchyma relationship is important and what he expect to find through the proposed studies using the laser capture microdissection:

> It is expected that the marked variations in epithelial-stromal ratio occurring during the various stages of breast development will influence the bidirectional connections between the components of cellular micro-environment (growth factors, hormones, and extracellular matrix) and the nucleus, leading to specific modifications in gene expression. The utilization of LCM in the present application will allow us to separate the parenchymal from the stromal components, thus allowing us to search for differential gene expression in the breast of the different groups under study.

Thus far the preliminary data has addressed the normal structure of the human breast. How this is related to cancer has yet to be explained and linked together. In this case, the applicant summarizes the previous work generated in his or her laboratory linking the importance of breast development and cancer. It is clear that the applicant has a gestalt of the whole process, but he or she cannot take for granted that the reviewers do as well, or leave them groping to discover it for themselves. To assume that the reviewers will search out missing information, or that they know what you know, is an arrogant position and it is not conducive to good science, or good writing for that matter. The following 300 words are devoted to making the connection between breast development and cancer. If additional material is needed it can be referred to in the appendix.

v. Breast development and the pathogenesis of breast cancer. The Lob 1, the most undifferentiated structure found in the breast of young nulliparous women, is the site of origin of ductal carcinomas[8,9] (see Appendix, Exhibit C). The finding that the most undifferentiated structures originated the most aggressive neoplasms support our hypothesis that the presence of Lob 1 explains the higher breast cancer risk of nulliparous women, since they represent the population with the highest concentration of undifferentiated structures in the breast.[25] Non-tumoral breast tissues from cancer-bearing lumpectomy or mastectomy specimens removed from nulliparous women have an architecture dominated by Lob 1; their overall architecture is similar to that of nulliparous females free of mammary pathology.[16,25] Although the breast tissues of parous women from the general population contain predominantly Lob 3 and a very low percentage of Lob 1, in those parous women that have developed breast cancer, their breast tissues also have Lob 1 as the predominant structure, appearing in this sense similar to those of nulliparous women.[27] It is of interest to note that all the parous breast cancer patients we have studied had a history of late first full-term pregnancy or familial history of breast cancer. The analysis of these samples allowed us to conclude that the architecture of the breast of parous women with breast cancer differs from that of parous women without cancer. The similarities found between the architecture of the breast of nulliparous women and that of parous women with cancer support our hypothesis that the degree of breast development is of importance in the susceptibility to carcinogenesis, and, furthermore, that parous women who develop breast cancer might exhibit a defective response to the differentiating influence of the hormones of pregnancy.[27]

In the previous five paragraphs the applicant has described the preliminary results that support the hypothesis to be developed. In the next part of the preliminary results, the applicant must provide data supporting the technical feasibility of that hypothesis. Once again, at the time the example application was written, most of the technology in question was in its nascent stage, yet it can apply to any other new methodologies. By the end of the proposal, the reviewers must not have any doubt that what was proposed is not only sound and novel, but feasible, and most importantly, that the applicant has the skills and experience to do it. In this case, it was important to demonstrate that the separation of the epithelia and the stroma was possible, and that enough material was obtained to retrieve good quality RNA that could be properly measured. For this purpose the applicant must summarize the work performed in his or her laboratory, in this case, work on the differential gene expression between Lob 1 of nulliparous and Lob 3 of parous premenopausal breast tissue, as described below:

a. **Laser capture microdissection (LCM).** The tissues used in this study were obtained from reduction mammoplasty specimens from three nulliparous premenopausal (28, 32 and 33 years of age) and three parous premenopausal women (26, 33 and 31 years of age). All the samples were frozen after removal within 10–20 minutes of the surgical procedure. The tissues were embedded in Tissue Tek OCT medium and frozen in liquid nitrogen. Eight-micron sections were obtained in cryostat and maintained at −80°C. The slides containing the frozen tissue were fixed in 70% ethanol for 30–40 seconds, stained with hematoxylin and eosin, followed by 5 second dehydration steps at 70, 95 and 100% and final dehydration step in xylene. Once air-dried the sections were laser microdissected with a PixCell II LCM system from Arcturus Engineering. Approximately 1×10^5 normal breast epithelial cells from Lob 1 and Lob 3 were obtained with an average of 50 to 60 captures per sample. The method is reliable in obtaining epithelial cells with 95% accuracy. There is small amount of material from the intralobular stroma that is difficult to separate. However, the material is free of the interlobular stroma.

b. **RNA quantification experiments.** We are able to reliably obtain total RNA from LCM-microdissected normal breast tissue; and the quality is adequate for subsequent enzymatic reactions. As to the quantity of RNA, we implemented two independent assays to measure low concentrations of RNA: fluorometric quantitation with the RNA dye Ribogreen™ and a linear RT-PCR assay developed in our laboratory for evaluating the expression level of the insulin growth factor binding protein 3 (IGF-BP3) gene. Based on these two assays, we estimated that approximately 40 ng of total RNA can be successfully isolated from 20 Lob 1 and 5 Lob 3 cross sections. This corresponds to approximately 50% of the theoretical recovery, suggesting that our procedures for tissue processing, LCM and RNA extraction approach ideal conditions. Based on these results, it appears that several hundreds ng of total RNA can be readily isolated upon LCM of lobular structures. However, it is necessary to implement procedures of RNA amplification in order to obtain enough RNA (20–40 μg) for microarray analysis. For this purpose, we have focused on RNA amplification protocols. In order to preserve signal fidelity it is critical that unavoidable artifacts are reduced to a minimum. Thus, in amplifying the small amounts of LCM-derived RNA, it is necessary to avoid unwanted bias and possible alterations in transcript representation that could be introduced as a consequence of the amplification step(s). It is particularly important to ensure adequate representation of low-abundance transcripts, which often carry considerable biological relevance. For these considerations, we avoided procedures of RNA amplification based on PCR techniques, since the exponential nature of the latter process is likely to introduce major artifacts in transcript abundance/representation. Rather we have implemented a protocol of the RNA polymerase of the bacteriophage T7, which affords linear amplification. The resultant amplified RNA generated with this protocol provides a proportional representation – both in size and complexity — of the starting material.[81] We obtained, based on the semi-quantitative reverse-transcription (RT-PCR) assay of the IGF-BP3, that the amplified RNA from the primary culture of cells obtained from Lob 3 provided a proportionally amplified RT-PCR signal in comparison to the starting material. Specifically, there was a linear relationship between the RNA concentration evaluated by the Ribogreen assay and the RTPCR signal intensity. By using this protocol, we have reproducibly obtained an approximately 5000-fold amplification in a single amplification step, when using high quality RNA (such as RNA extracted from the primary cell culture of Lob 3). When using RNA of poorer quality, such as RNA extracted from LCM-microdissected tissue, the amplifications achieved in a single step was 50–100-fold. This renders necessary two successive amplification steps in order to obtain a 5000-fold amplification. In our hands, a third amplification step, as reported in the literature, is unnecessary. In conclusion, only two amplification steps of LCM-extracted RNA are necessary, thus minimizing unwanted bias and possible alterations in transcript representation, and ensuring adequate representation of low-abundance transcripts.

In the next section the applicant should provide further proof that the technique works and that reliable data can be obtained. Here it is important to note that the majority of grant applications are requesting government funds. What this means is that the applicant is seeking to spend the "people's money," therefore the applicant must not only be qualified and the idea sound, but he or she needs to offer the assurance that the funds provided to him or her to perform the research is a secure investment that will generate future data that will be part of the common human knowledge on the subject.

The next paragraph describes such data:

c. **Comparative analysis of gene expression between breast epithelial cells from Lob 1 and Lob 3.** This study was performed using the RNA obtained from microdissected Lob 1 and Lob 3 of three nulliparous and three parous premenopausal women using LCM (see section 3.vi.a), and hybridized to cDNA array membranes that contain 1,176 human genes (Clontech Human Cancer 1, 2 array). Lob 3 of parous women showed 82 genes differentially overexpressed or downregulated when compared with those expressed in Lob 1. With this array it is clear that Lob 3 has a gene expression profile significantly different from that of lobules type 1 of the nulliparous breast.

Following these data the applicant raises an important concern, which is the functional role of the genes identified:

Although it is difficult to discuss in detail in this application the possible functional role of each one of these genes, it is of interest to note that the RhoE gene is amplified 14 folds in the Lob 3 of the parous breast. This gene belongs to a small G protein superfamily that consists of the Ras, Rho, Rab, Arf, Sar1, and Ran families.[12,59,61,82–84] *In vivo*, RhoE is found exclusively in the GTP-bound form, suggesting that unlike previously characterized small GTPases, RhoE may be normally maintained in an activated state.[83] This could be an important function, considering that, this gene might remain expressed even after involution of Lob 3 to Lob 1 in the postmenopausal state. The other gene that is significantly overexpressed (five-fold) in the Lob 3 is the protein tyrosine phosphatase

(PPTPCAAX1 or PRL-1). This gene encodes a unique nuclear protein-tyrosine phosphatase,[85] which is regulated by a mechanism different from those of other immediate-early genes such as c-fos and c-jun.[86] This gene has been shown to be upregulated in villus, but not crypt enterocytes, and in confluent differentiated but not undifferentiated proliferating Caco-2 colon carcinoma cells.[87] In other systems it is also related to differentiation,[88] development and regeneration.[89] Therefore its function in the breast epithelial cells of Lob 3 may be related to a differentiation role more than a proliferative related process, because these lobules have a lower proliferative activity than Lob 1 (insert supporting figure here). Insulin-like growth factor binding protein-3 (IGFBP-3) is significantly overexpressed in Lob 3 when compared with Lob 1 of the breast of pre-menopausal women. This gene codes a specific binding protein for the insulin-like growth factors. IGFBP-3 modulates the mitogenic and metabolic effects of IGFs, and form a ternary complex with IGF-I or IGF-II and a 85-kd glycoprotein acid-labile subunit ALS.[90] IGFBP-3 may also play more active, IGF-independent roles in growth regulation of cancer cells.[91] IGFBP-3 protein levels are developmentally regulated and influenced by a number of hormonal stimuli both *in vitro* and *in vivo*.[92] P53 may regulate apoptosis in tumor cells via transactivation of IGFBP-3 gene.[93]

The applicant should then provide further support to the preliminary data by introducing additional data, that although they are not shown, conceptually support the point.

> We have found (data not shown) that IGBP-3 could be modulated by hCG and be one important pathway in the differentiation effect of this hormone in the mammary gland. Its expression in the Lob 3 is a new finding that requires further investigation.

The applicant should end this portion of the preliminary results by explaining the physiological role of the changes found, as well as the ones expected to be found in the study:

> The final biological significance of the genes found in the process of differentiation of the breast, the persistence of their expression during

postmenopausal involution and how they are regulated by the reproductive history of the woman are not known. There is no clear understanding either how these or other genes act in the protection induced by pregnancy. Of interest was the observation that there was no difference between Lob 1 and Lob 3 in the expression of cytoskeletal proteins such as cytokeratins 18, 19, and 8, that are overexpressed in tumor cells.[93] Known genes that are overexpressed in breast cancer, such as HER-2/neu[38] and mucin[39,40] were not expressed in any of the lobular structures. Other genes, such as fibroblast-like growth factor 1 (FGF-1),[38] insulin growth factor-1 (IGF-1) binding protein-2,[94,95] and Zinc-α-2-glycoprotein, that are generally up or down regulated in the neoplastic process, were not differentially expressed in the Lob 3 vs. Lob 1. In addition, cells derived from the differentiated Lob 3 are resistant to grow *in vitro* and do not express transformation phenotypes upon carcinogen treatment, as cells from Lob 1 do.[13,14,21,22] We have confirmed the differential expression of some of those genes that were overexpressed in Lob 3 using semiquantitative RT-PCR. The results of the RT-PCR might not quantitatively reflect the fold difference found by microarray, but they agree with the results. For example PRL-1, RhoE and IGFBP 3 were all overexpressed using both procedures. Beta actin was equally expressed using both techniques.

So far, the applicant will have used the eight pages allotted to describe the preliminary results, and provided support for the hypothesis to be tested. However, it is important to summarize and provide a conceptual framework that the reviewers can use when writing their critiques. I strongly advise a summary of this preliminary data section that highlights aspects to be remembered. For this application, under the heading Unifying Concepts, I decided to write the following summary:

Unifying concepts. Breast cancer originates in undifferentiated terminal structures of the mammary gland. The terminal ducts of the Lob 1 of the human female breast, which are the sites of origin of ductal carcinomas, are at their peak of cell replication during early adulthood, a period during which the breast is more susceptible to carcinogenesis. The susceptibility of Lob 1 to undergo neoplastic transformation has been confirmed by *in vitro* studies, which have shown that this structure has the highest proliferative activity and rate of carcinogen binding to the DNA. More importantly, when treated with carcinogens *in vitro* its epithelial cells express phenotypes indicative of cell transformation.[13,14] These studies indicate that in the human breast the target cell of carcinogens is found in a specific compartment whose characteristics are the determinant factors in the initiation event. These target cells will become the stem cells of the neoplastic event, depending upon: (a) topographic location within the mammary gland tree, (b) age at exposure to a known or putative genotoxic agent, and (c) reproductive history of the host. The higher incidence of breast cancer observed in nulliparous women supports this concept, because it parallels the higher cancer incidence elicited by carcinogens in rodents when exposure occurs at a young age (see exhibit C in Appendix). In addition, it has been shown that early parity is associated with a pronounced decrease in the risk of breast cancer an additional live births confer greater risk reduction at an approximate rate of 10% per rate.[24] Thus, the protection afforded by early full-term pregnancy in women could be explained by the higher degree of differentiation of the mammary gland at the time at which an etiologic agent or agents act. Even though differentiation significantly reduces cell proliferation in the mammary gland, the mammary epithelium remains capable of responding with proliferation to given stimuli, such as a new pregnancy. Under these circumstances, however, the cells that are stimulated to proliferate are from structures that have already been primed by the first cycle of differentiation, thus creating a second type of stem cells that are able to metabolize the carcinogen and repair the induced DNA damage more efficiently than the cells of the virginal gland, and are less susceptible to carcinogenesis, as it has been demonstrated in the rodent experimental system.[76] However, a carcinogenic stimulus powerful enough may overburden the system, thereby initiating successfully a neoplastic process. Such conditions may explain the small fraction of women developing breast cancer after an early first full-term pregnancy, i.e. after completion of the first cycle of differentiation. The relevance of our work lies in the vis-à-vis comparison of *in vivo* and *in vitro* studies of the human breast that validate experimental data for extrapolation to the human situation. The finding that differentiation is a powerful inhibitor of cancer initiation provides a strong rationale for identifying the genes that control this process. The knowledge gained will provide novel tools for developing rational strategies for breast cancer prevention.

If the granting agency allows an appendix section it could be used to provide additional material to the reviewers. If the applicant chooses to include an appendix, the summary paragraph should make reference to the additional information provided therein.

Experimental Design and Methods

For this final section of the grant application, it is convenient to begin with the general and gradually move towards the particular. For example, describe the aims followed by a brief summary of how they will be accomplished. After that, start describing how the experiment or experiments will be performed, or how the samples will be studied. The statistical portion should be at the end and must be seriously considered, never overlooked or underestimated.

Specific Procedures for Specific Aim 1: To characterize the specific gene expression profile of women at "low" and "high" risk of developing breast cancer because of reproductive history. Gene expression in normal breast tissue obtained from postmenopausal women with a history of one or more early full term pregnancies will be compared to gene expression in normal breast tissue from postmenopausal women who are nulliparous. For this purpose, we will use a large-scale analysis of gene expression using available human cDNA libraries for determining the expression pattern of known and unknown (ESTs) genes in normal breast tissues obtained from postmenopausal women with a history of one or more early full term pregnancies by comparison to those expressed in the "high breast cancer risk" population, represented by nulliparous women. We will confirm the gene expression of the identified clusters of genes by utilizing real-time quantitative RT-PCR and *in situ* hybridization or immunocytochemistry, where antibodies for the encoded gene protein product are available, on constructed arrays of paraffin embedded tissues of the samples included in this study.

A critical aspect in the methodology dealing with human material is the sample collection. For this reason this portion needs to be carefully considered, as exemplified below:

> **i. Sample Collection.** This study will be carried out using normal human breast tissue obtained from postmenopausal women who will be divided into two groups, according to their reproductive history). As a "reference standard" we will utilize mRNA from Lob 3 and Lob 1 and from human breast epithelial cell lines. The first group (Group 1) representing the "low risk group" is composed of postmenopausal women without breast cancer who completed their first full term pregnancy (FFTP) before age 24. Group 2, are postmenopausal women without breast cancer who are nulliparous. Our "reference standard" is mRNA obtained from nulliparous and parous women, 15 samples each, and human breast epithelial cells lines available in our laboratory such as, MCF10F, MCF10A, MCF12A, MCF7, MDA-MB-234m, T24, B20, and 76N (100–102), mixed at equal ratios.
>
> The selection of postmenopausal women, ranging in age from 50 to 65 years of age, has the dual advantage of avoiding the potential bias introduced by cyclic hormonal variations in premenopausal women. Postmenopausal is defined as at least one year since last menses if menopause occurred naturally or basal serum follicle stimulating hormone (FSH) >40 ng/ml if menopause was surgical and the participant is <60 years old.

The source of the human sample collection must be thoroughly described and adequate approval from the Internal Review Board (IRB), must be procured. Examples of the details required are described below:

> The breast samples have been collected and they are already available in our tissue bank. They consist of 60 breast samples obtained from 30 white females, 28 black females, 1 Hispanic, and 1 of unknown racial origin. For this purpose the tissues were fixed for light microscopy, and a portion of each tissue was snap-frozen at the time of collection. From these collected samples we have the following information: age, race, menopausal status, age at menarche, day of last menstruation, number of miscarriages, year of first full term pregnancy, number of pregnancies. None of these women

have a familial history of breast cancer, received replacement therapy or previously underwent surgical procedures for benign or malignant diseases of the breast. For specific aim 1 we will collect new samples. The numbers of samples that are expected to be collected are based on estimation of three years accrual. Based on these data we are expecting to obtain 17 samples per year that will fulfill our criteria. Each of these samples will be associated with it clinical information obtained from the participant. A study coordinator will be in charge of transporting samples, logging, banking and coordinating all the pertinent work that will be performed for each sample.

As indicated, the statistical approach is vital in every application. I have transcribed, for educational purposes, the statistical analysis used in this specific application.

Statistical Approach. Univariate analyses will be used to select differentially expressed genes for inclusion on the custom arrays. Multivariate analyses will be conducted to identify gene clusters sharing similar expression profiles.

This aim will be conducted in three stages:

Stage 1. The purpose of the first stage is to examine the candidate genes for differential expression in Groups 1 and 2. This will be based on normalized data from 60 arrays (15 arrays for each of the epithelial and stromal components from each of groups 1 and 2) with a total of 40,000 candidate genes per array set. Specifically, the goal is to identify up to 4,200 genes for further evaluation using custom arrays in stage 2. In order to obtain representative parenchymal tissue (Lob 1) and stromal cells from the intralobular space and interlobular portion of the breast tissue, LCM will be performed in 15 samples each of Groups 1 and 2. Total RNA obtained from 500–1000 LCM-microdissected cells from each sample, will be subjected to 3 cycles of "linear" amplification by *in situ* RT-RNA-PCR and then used to carry out gene expression analysis.

Stage 2. The set of up to 4,200 known genes and ESTs that were selected in stage 1 will be spotted on replicate slides at the FCCC Microarray facility. Genes such as β-actin, α-tubulin, cyclophilin, keratin 8, 9 and 19 that have not been affected in our previous studies comparing Lob1 vs. Lob 3, as well as other housekeeping genes, will be included in the array as controls for the purpose of scaling and normalization. RNA from mammary epithelial and stromal cells will be extracted and amplified as described above from at least 45 samples from each of the two risk groups identified in the previous section (insert supporting data here). The aRNA will be used to synthesize cDNA probes to screen replicate array slides. Each hybridization will be repeated twice with dye-flip. The gene expression profiles obtained from each hybridization will be normalized to decrease variations attributable to differences in overall mRNA abundance, cDNA probe synthesis and hybridization efficiency.

A general description of quality control issues that could be raised is an important aspect of the Methods and Procedures section:

Quality Control. As part of the quality control in the reliability of the hybridization signal and experimental variables unrelated to the differences in hybridization probes, the array will be hybridized independently three times with Cy3 and Cy5 labeled liver cDNA probes simultaneously. It is expected that not less than 80% of the clones will exhibit a Cy5 to Cy3 signal ratio of less than 1.5. A scatter plot can be constructed with the values of the Cy5/Cy3 ratio; a value of near 1 is expected. In the hybridization results of the microarray with Cy3-labeled cDNA from the nulliparous breast tissue and Cy5-labeled cDNA probe from the breast tissue of parous women (FFTP), it will be expected that at least 30% of the cDNAs will exhibit more than two-fold change in expression level and about 10% of the cDNAs will have a difference in expression greater than three-fold. The scatter plot must reveal in these cases a very wide distribution pattern.

Stage 3. Validation of the differential gene expression pattern between the "low" and "high breast cancer risk" populations. For this purpose the status of genes found to exhibit differential expression patterns associated with different risk levels for breast cancer development will be further confirmed by alternative methods. Such methods will include QRT-PCR amplification using RNA preparations that will have been tested by expression microarray analysis (Specific Aim 1, stage 2), as well as by *in situ* hybridization in tissue array. One third of the same RNA pool used for the gene array hybridizations will be reverse-transcribed using 50 μg/ml oligo (dT), 500 μM deoxynucleotide triphophosphate, and 200 units of Superscript II reverse transcriptase (Life Technologies) for 1 hour at 37°C, and the resulting first strand cDNA diluted and used as template for **Real-time Quantitative PCR (QRT-PCR)** analysis. Sequences for genes identified using array technology will be determined by direct sequence analysis and confirmed using NCBI GenBank and UniGene databases. The specificity of amplicon sequence selection will be determined using two methods. First, primer and probe sequences that specifically detect the experimental gene sequence, as determined by means of the NCBI Blast module, will be utilized. Second, amplicons generated during the PCR reaction will be analyzed using the first derivative primer melting curve software supplied together with the SmartCycler system from Cepheid. Analysis of gene expression will be carried out using a dual-module SmartCycler system with 32-sample capacity and four-color optical capability, which is currently available at FCCC. The SmartCycler System software optically monitors each reaction site as the fluorescent signals develop. Growth curves are displayed in real time as amplification occurs, and the presence of amplified product can be confirmed when the fluorescent signal exceeds a defined threshold. By using multiple fluorescent reporter dyes, up to four targets can be detected simultaneously in a single reaction mixture. A relative standard curve representing four four-fold dilutions of breast stock cDNA (1:2.5, 1:10, 1:40, and 1:160) will be used for linear regression analysis of unknown samples. The expression of a number of housekeeping genes, including cyclophilin 33A and GAPDH, will be used to normalize for variances in input cDNA.

In the example provided above, the main consideration is sample handling and methods of analysis rather than a detailed protocol of, for example, the immunoytochemistry or how specific reagents will be used. The message that the applicant must provide to the reviewers is how the analysis of the data will be done. In addition to this, at the end of each specific aim is an important paragraph designated *Expected Results and Significance* of the work proposed, as described below:

> **Expected Results and Significance.** The methodology proposed to be employed for this Aim will allow us to test the hypothesis that early parity induces activation of specific sets of genes leading to cell differentiation. Although we do not know which will be the final pattern of gene expression, we postulate, based on our preliminary results, that genes controlling cell proliferation, programmed cell death and DNA repair would be involved. The identification of genes that are induced by pregnancy is crucial for a better understanding of the process of gland differentiation, and therefore breast cancer prevention. This can open new avenues for exploring the effect of physiological mechanisms, such as pregnancy and hormonally induced differentiation in the therapy of breast cancer. Our hypothesis is that pregnancy produces permanent genomic changes in the mammary epithelial cells that are the basis for the protective effect of pregnancy after the phenotypic expression of glandular differentiation is completed. The pattern of gene expression could potentially be used as useful intermediate end points for evaluating the degree of mammary gland differentiation and for assessing its potential risk for neoplastic transformation.

This paragraph ends the specific procedures for Aim 1. The same format should be followed with all subsequent aims.

The other sections of the application, such as References, must also be carefully monitored for accuracy. Additional administrative information is also required in grant applications. The example provided in this chapter is only the core of a grant application, the most difficult, and most important to learn and master.

As part of this section, a **timetable** is also recommended. This timetable establishes the landmarks to be achieved during the grant period. It also serves to justify the time requested for performing the project. If the grant is requested for a period of three years but the timetable reflects five years worth of work, the reviewers will surely notice the discrepancy and this could negatively affect the application. The sense of timing, or tempo, in experimentation is also a quality that the reviewers will appreciate in the applicant. Although development of a budget is beyond the scope of this book, it is important that each item be properly justified and requested with each aim and the timeline in mind.

Finally, writing grant applications is a never-ending experience and probably the most challenging one that the scientific apprentice will have to face. It is a continuous exercise that requires the right mental disposition and training. In this respect the scientific apprentice is like a marathoner: only through training and practice can progress be made.

References

1. Greenle RT, Murray T, Boldin S and Wingo P. Cancer Statistics 2000. *CA Cancer J Clin* 50-7-23, 2000.
2. Henderson BE, Ross RK and Pike MD. Hormonal chemoprevention of cancer in women. *Science* 259: 6n–M8, 1993.
3. MacMahon B, Cole P, Lin TM *et al*. Age at first birth and breast cancer risk. *Bull Nat'l Hlth Org* 43:209, 1970.
4. Vessey MD, McPherson K, Roberts MM, Neil A and Jones L. Fertility and the risk of breast cancer. *Br J Cancer* 52:625–628, 1985.
5. Trapido EJ. Age at first birth, parity and breast cancer risk. *Cancer* 51:946–948, 1983.
6. Russo J, Tay LK and Russo IH. Differentiation of the mammary gland and susceptibility to carcinogenesis: A Review. *Breast Cancer Res Treat* 2:5–73, 1982.
7. De Waard F and Trichopoulos D. A unifying concept of the etiology of breast cancer. *Int J Cancer* 41:666–669, 1988.
8. Russo J, Gusterson BA, Rogers AE, Russo IH, Welling SR and Van Zwieten MJ. Comparative study of human and rat mammary tumorigenesis. *Lab Invest* 62:244–278, 1990.
9. Wellings SR, Jansen MM and Marcum RG. An atlas of sub-gross pathology of the human breast with special reference to possible pre-cancerous lesions. *JNCI* 55:231–275, 1975.
10. Russo J and Russo IH. Influence of differentiation and cell kinetics on the susceptibility of the mammary gland to carcinogenesis. *Cancer Res* 40:2677–2687, 1980.

11. Tay LK and Russo J. 7,12-dimethylbenz(a)anthracene (DMBA) induced DNA binding and repair synthesis in susceptible and non-susceptible mammary epithelial cells in culture. *J Natl Cancer Inst* 67:155–161, 1981.

12. Tay LK and Russo J. Formation and removal of 7,12-dimethylbenz(a)anthracene-nucleic acid adducts in rat mammary epithelial cells with different susceptibility to carcinogenesis. *Carcinogenesis* 2:1327–1333, 1981.

13. Russo J, Reina D, Frederick J and Russo IH. Expression of phenotypical changes by human breast epithelial cells treated with carcinogens *in vitro*. *Cancer Res* 48:2837–2857, 1988.

14. Russo J, Calaf G and Russo IH. A critical approach to the malignant transformation of human breast epithelial cells. *CRC Critical Reviews in Oncogenesis* 4:403–417, 1993.

15. Russo J and Russo IH. Toward a physiological approach to breast cancer prevention. *Cancer Epidem Biomar Prev* 3:353–364, 1994.

16. Russo J and Russo IH. Differentiation and Breast Cancer Development. In *Advances in Oncobiology*, Vol. 2, Heppner G (Ed.), JAI. Press, Inc., pp. 1–10, 1998.

17. Russo J and Russo IH. Toward a Unified Concept of Mammary Tumorigenesis. *Prog Clin Biol Res* 396:1–16, 1997.

18. Russo J, Ao X, Grill C and Russo IH. Pattern of distribution of cells positive for estrogen receptor α and progesterone receptor in relation to proliferating cells in the mammary gland. *Breast Cancer Res Treat* 53:217–227, 1999.

19. Hu YF, Silva IDCG, Russo IH, Ao X and Russo J. A novel serpin gene cloned from differentiated human breast epithelial cells is a potential tumor suppressor. *Proc Am Assoc Cancer Res* 39:114, 1998.

20. Hu YF, Russo IH, Ao X and Russo J. Mammary-derived growth inhibitor (MDGI) cloned from human breast epithelial cells is expressed in fully differentiated lobular structure. *Int J Oncol* 11:5–11, 1997.

21. Russo J and Russo IH. Development of the human mammary gland. In *The Mammary Gland*, Neville MC and Daniel C (Eds.). Plenum Publishing, Inc., New York, pp. 67–93, 1987.

22. Russo J, Rivera R and Russo J. Influence of age and parity on the development of the human breast. *Breast Cancer Res Treat* 23:211–218, 1992.

23. Russo J and Russo IH. Biological and molecular bases of mammary carcinogenesis. *Lab Invest* 57:112–137, 1987.

24. Lambe M, Hsieh C-C, Chan H-W, Ekbom A, Trichopoulos D and Adami Parity HO. Age at first and last birth, and risk of breast cancer: A population-based study in Sweden. *Breast Cancer Res Treat* 38:305–311, 1996.
25. Russo J and Russo IH. The cellular basis of breast cancer susceptibility. *Oncol Res*, 11:169–178, 1999.
26. Russo J and Russo IH. Development pattern of human breast and susceptibility to carcinogenesis. *Eur J Cancer Preven* 2:85–100, 1993.
27. Russo J, Romero AL and Russo IH. Architectural pattern of the normal and cancerous breast under the influence of parity. *J Cancer Epidem Biom Prev* 3:219–224, 1994.
28. Russo IH and Russo J. Developmental stage of the rat mammary gland as determinant of its susceptibility to 7,12-dimethylbenz(a)anthracene. *J Natl Cancer Inst* 61:1439–1449, 1978.
29. Hancock SL, Tucker MA and Hoppe RT. Breast cancer after treatment of Hodgkin's disease. *J Natl Cancer Inst* 85:25–31, 1993.
30. Russo J and Russo IH. DNA labeling index and structure of the rat mammary gland as determinants of its susceptibility to carcinogenesis. *J Natl Cancer Inst* 61:1451–1459, 1978.
31. Russo J, Wilgus G and Russo IH. Susceptibility of the mammary gland to carcinogenesis. Differentiation of the mammary gland as determinant of tumor incidence and type of lesion. *Am J Pathol* 96:721–736, 1979.
32. Russo J. Basis of cellular autonomy in the susceptibility to carcinogenesis. *Toxicologic Pathology* 11:149–166, 1983.
33. Russo J and Russo IH. Role of differentiation on transformation of human epithelial cells. In *Cellular and Molecular Biology of Mammary Cancer*, Medina D et al. (Eds.). Plenum Press: New York, pp. 399–417, 1987b.
34. Dabelow A. Die Milchdruse. In *Handbuch der Mikroskopischen Anatomic des Menschen*, Bargmann W (Ed.). Springer-Velag: Berlin, pp. 277–485, 1957.
35. Salazar H and Tobon H. Morphologic changes of the mammary gland during development, pregnancy and lactation. In *Lactogenic Hormones, Fetal Nutrition and Lactation*, Josimovich, J. (Ed.). Wiley, New York, pp. 221–277, 1974.
36. Vorherr H. *The Breast*, Academic Press: New York, pp. 1–18, 1974.
37. Russo J and Russo IH. Development of the Human Breast. In *Encyclopedia of Reproduction* Knobil E and Neill JD (Eds.). Academic Press: New York, Vol. 3, pp. 71–80, 1998.

38. Welch DR and Wei LL. Genetic and epigenetic regulation of human breast cancer progression and metastasis. *Endocrine Related Cancer* 5:155–197, 1998.

39. Aoki R, Tanaka S, Haruma K, Yoshihara M, Sumii K *et al.* MUC-1 expression as a predictor of the curative endoscopic treatment of submucosally invasive colorectal carcinoma. *Dis Colon Rectum* 41:1262–1272, 1998.

40. Segal Eiras A and Croce MV. Breast cancer associated mucin: A review. *Allergol. Immunopathol* 25:176–181, 1997.

41. Russo J, Saby J, Isenberg W and Russo IH. Pathogenesis of mammary carcinomas induced in rats by 7,12-dimethylbenz(a)anthracene. *J Natl Cancer Inst* 59:435–445, 1977.

42. Russo J and Russo IH. Is differentiation the answer in breast cancer prevention? *IRCS Med Sci* 10:935–941, 1982.

43. Russo J, Mills MJ, Moussall MJ and Russo IH. Influence of human breast development on the growth properties of primary cultures. *In Vitro Cell Develop Biol* 25:643–649, 1989.

44. DeRisi J, Penland L, Brown PO, Bittner ML, Meltzer PS, Ray M, Chen Y, Su YA and Trent JM. Use of a cDNA microarray to analyse gene expression patterns in human cancer. *Nat Genet* 14:457–60, 1996.

45. Heller RA, Schena M, Chai A, Shalon D, Bedilion T, Gilmore J, Woolley DE and Davis RW. Discovery and analysis of inflammatory disease-related genes using cDNA microarrays. *Proc Natl Acad Sci USA* 94:2150–2155, 1997.

46. Schena M, Shalon D, Davis RW and Brown PO. Quantitative monitoring of gene expression patterns with a complementary DNA microarray. *Science* 270:467–470, 1995.

47. Schena M, Shalon D, Heller R, Chai A, Brown PO and Davis RW. Parallel human genome analysis: Microarray-based expression monitoring of 1000 genes. *Proc Natl Acad Sci USA* 93:10614–10619, 1996.

48. Ramsay G. DNA chips: State-of-the art. *Nat Biotechnol* 16:40–44, 1998.

49. Schena M, Heller RA, Theriault TP, Lachenmeier E and Davis RV. Microarrays: Biotechnology's discovery platform for functional genomics. *Trend Biotechnol* 16:301–316, 1998.

50. Eisen MB, Spellman PT, Brown PO and Botstein D. Cluster analysis and display of genome-wide expression patterns. *Proc Natl Acad Sci USA* 95:14863–14868, 1998.

51. Spellman PT, Sherlock G, Zhang MQ, Iyer VR, Anders K, Eisen MB, Brown PO, Botstein D and Futcher B. Comprehensive identification of cell

cycle-regulated genes of the yeast *Saccharomyces cerevisiae* by microarray hybridization. *Mol Biol Cell* 9:3273–3297, 1998.

52. Iyer VR, Eisen MB, Ross DT, Schuler G, Moore T, Lee JCF, Trent JM, Staudt LM, Hudson J, Jr., Boguski MS, Lashkari D, Shalon D, Botstein D and Brown PO. The transcriptional program in the response of human fibroblasts to serum (see comments). *Science*, 283:83–87, 1999.

53. Golub TR, Slonim DK, Tamayo P, Huard C, Gaasenbeek M, Mesirov JP, Coller H, Loh ML, Downing JR, Caligiuri MA, Bloomfield CD and Lander ES. Molecular classification of cancer: class discovery and class prediction by gene expression monitoring. *Science* 286:531–537, 1999.

54. Perou CM, Sorlie T, Eisen MB, van de Rijn M, Jeffrey SS, Rees CA, Pollack JR, Ross DT, Johnsen H, Akslen LA, Fluge O, Pergamenschikov A, Williams C, Zhu SX, Lonning PE, Borresen-Dale AL, Brown PO and Botstein D. Molecular portraits of human breast tumours. *Nature* 406:747–752, 2000.

55. Ross DT, Scherf U, Eisen MB, Perou CM, Rees C, Spellman P, Iyer V, Jeffrey SS, Van de Rijn M, Waltham M, Pergamenschikov A, Lee JC, Lashkari D, Shalon D, Myers TG, Weinstein JN, Botstein D and Brown PO. Systematic variation in gene expression patterns in human cancer cell lines. *Nat Genet* 24:227–235, 2000.

56. Scherf U, Ross DT, Waltham M, Smith LH, Lee JK, Tanabe L, Kohn KW, Reinhold WC, Myers TG, Andrews DT, Scudiero DA, Eisen MB, Sausville EA, Pommier Y, Botstein D, Brown PO and Weinstein JN. A gene expression database for the molecular pharmacology of cancer. *Nat Genet* 24:236–244, 2000.

57. Hagberg G. From magnetic resonance spectroscopy to classification of tumors. A review of pattern recognition methods. *NMR Biomed* 11:148–156, 1998.

58. Ochs MF, Stoyanova RS, Arias-Mendoza F and Brown TR. A new method for spectral decomposition using a bilinear Bayesian approach. *J Magn Reson* 137:161–176, 1999.

59. Saeed N. Magnetic resonance image segmentation using pattern recognition, and applied to image registration and quantitation. *NMR Biomed* 11:157–167, 1998.

60. Stoyanova R, Kuesel AC and Brown TR. Application of principal component analysis for NMR spectral quantitation. *J Magn Reson Ser A* 115:265–269, 1995.

61. Brown PO and Botstein D. Exploring the new world of the genome with DNA microarray. *Nat Genet* 21(Suppl.):33–37, 1999.

62. Stutz J and Cheesman P AutoClass — A Bayesian approach to classification. In *Maximum Entropy and Bayesian Methods*, Skilling J and Sibisi S (Eds.) Kluwer Academic Publishers: Dordrecht, 1995.

63. Russo J and Russo IH. Role of hormones in human breast development. The menopausal breast. In *Progress in the Management of Menopause*. Parthenon Publishing: London, pp. 184–193, 1997.

64. Russo J and Russo IH. Role of differentiation in the pathogenesis and prevention of breast cancer. *Endocrine Related Cancer* 4:7–21, 1997.

65. Kumar V, Stack GS, Berry M, Jin JR and Chambon P. Functional domains of the human estrogen receptor. *Cell* 51:941–951, 1987.

66. King RJB. Effects of steroid hormones and related compounds on gene transcription. *Clin Endocrinol* 36:1–14, 1992.

67. Soto AM and Sonnenschein C. Cell proliferation of estrogen-sensitive cells-the case for negative control. *Endocr Rev* 48:52–58, 1987.

68. Huseby RA, Maloney TM and McGrath CM. Evidence for a direct growth-stimulating effect of estradiol on human MCF-7 cells *in vitro*. *Cancer Res* 144:2654–2659, 1987.

69. Huff KK, Knabbe C, Lindsey R, Kaufman D, Bronzert D, Lippman ME and Dickson RB. Multihormonal regulation of insulin-like growth factor-1-related protein in MCF-7 human breast cancer cells. *Mol Endocrinol* 2:200–208, 1988.

70. Dickson RB, Huff KK, Spencer EM and Lippman ME. Introduction of epidermal growth factor related polypeptides by 17b-estradiol in MCF-7 human breast cancer cells. *Endocrinol* 118:138–142, 1986.

71. Page MJ, Field JK, Everett P and Green CD. Serum regulation of the estrogen responsiveness of the human breast cancer cell line MCF-7. *Cancer Res* 43:1244–1250, 1983.

72. Katzenellenbogen BS, Kendra KL, Norman MJ and Berthois Y. Proliferation, hormonal responsiveness and estrogen receptor content of MCF-7 human breast cancer cells growth in the short-term and long-term absence of estrogens. *Cancer Res* 47:4355–4360, 1987.

73. Aakvaag A, Utaacker E, Thorsen T, Lea OA and Lahooti H. Growth control of human mammary cancer cells (MCF-7 cells) in culture: Effect of estradiol and growth factors in serum containing medium. *Cancer Res* 50:7806–7810, 1990.

74. Dell'aquila ML, Pigott DA, Bonaquist DL and Gaffney EV. A factor from plasma derived human serum that inhibits the growth of the mammary cell

line MCF-7 characterization and purification. *J Natl Cancer Inst* 72:291–298, 1984.

75. Russo IH and Russo J. Role of hormones in cancer initiation and progression. *J Mammary Gland Biol Neoplasia* 3:49–61, 1997.

76. Russo IH and Russo J. Mammary gland neoplasia in long-term rodent studies. *Environ Health Perspect* 104:938–967, 1996.

77. Russo IH, Koszalka M, Russo J. Human chorionic gonadotropin and rat mammary cancer prevention. *J Natl Cancer Inst* 82:1286–1289, 1990.

78. Russo IH, Koszalka M, Russo J. Comparative study of the influence of pregnancy and hormonal treatment on mammary carcinogenesis. *Br J Cancer* 64:481–484, 1991.

79. Xie J and Haslam SZ. Extracellular matrix regulates ovarian hormone-dependent proliferation of mouse mammary epithelial cells. *Endocrinology* 138(6):2466–2473, 1997.

80. Petersen OW, Ronnov-Jessen L, Weaver VM and Bissell MJ. Differentiation and cancer in the mammary gland: Shedding light on an old dichotomy. *Adv Cancer Res* 75:135–161, 1998.

81. Kacharmina JE, Crino PB and Eberwine J. Methods in Enzymology 303:3–18, 1999.

82. Hall A. Rho GTPases and the actin cytoskeleton. *Science* 279(5350):509–514, 1998.

83. Foster R, Hu KQ, Lu Y, Nolan KM, Thissen J and Settleman J. Identification of a novel human Rho protein with unusual properties: GTPase deficiency and *in vivo* farnesylation. *Mol Cell Biol* 6:2689–2699, 1996.

84. Guasch RM, Scambler P, Jones GE and Ridley AJ. RhoE regulates actin cytoskeleton organization and cell migration. *Mol Cell Biol* 18(8):4761–4771, 1998.

85. Peng Y, Du K, Ramirez S, Diamond RH and Taub R. Mitogenic up-regulation of the PRL-1 protein-tyrosine phosphatase gene by Egr-1. Egr-1 activation is an early event in liver regeneration. *J Biol Chem* 274(8):4513–4520, 1999.

86. Takano S, Fukuyama H, Fukumoto M, Kimura J, Xue JH, Ohashi H and Fujita J. PRL-1, a protein tyrosine phosphatase, is expressed in neurons and oligodendrocytes in the brain and induced in the cerebral cortex following transient forebrain ischemia. *Brain Res Mol Brain Res* 40(1):105–115, 1996.

87. Diamond RH, Peters C, Jung SP, Greenbaum LE, Haber BA, Silberg DG, Traber PG and Taub R. Expression of PRL-1 nuclear PTPase is associated

with proliferation in liver but with differentiation in intestine. *Am J Physiol* 271(1 Pt 1):G121–129, 1996.

88. Rundle CH and Kappen C. Developmental expression of the murine Prl-1 protein tyrosine phosphatase gene. *J Exp Zool* 283(6):612–617, 1999.

89. Peng Y, Genin A, Spinner NB, Diamond RH and Taub R. The gene encoding human nuclear protein tyrosine phosphatase, PRL-1. Cloning, chromosomal localization, and identification of an intron enhancer. *J Biol Chem* 273(27):17286–17295, 1998.

90. Phillips LS, Pao CI and Villafuerte BC. Molecular regulation of insulin-like growth factor-I and its principal binding protein, IGFBP-3. *Prog Nucleic Acid Res Mol Biol* 60:195–265, 1998.

91. Oh Y. IGFBPs and neoplastic models. New concepts for roles of IGFBPs in regulation of cancer cell growth. *Endocrine* 7(1):111–113, 1997.

92. Cubbage ML, Suwanichkul A and Powell DR. Insulin-like growth factor binding protein-3. Organization of the human chromosomal gene and demonstration of promoter activity. *J Biol Chem* 265(21):12642–12649, 1990.

93. Coverley JA and Baxter RC. Phosphorylation of insulin-like growth factor binding proteins. *Mol Cell Endocrinol* 128(1–2):1–5, 1997.

94. Brotherick I, Robson CN, Bronell DA, *et al*. Cytokeratin expression in breast cancer: Phenotypic changes associated with disease progression. *Cytometry* 32:301–308, 1998.

95. Manni, A, Badger B, Wei L, *et al*. Hormonal regulation of insulin-growth factor II and insulin growth factor binding protein expression by breast cancer cells *in vivo*. Evidence for epithelial stromal interactions. *Cancer Research* 54:2934–2942, 1994.

Chapter 5

What is a Research Laboratory?

Definition of a Research Laboratory

While the laboratory is a physical workplace, the room or series of rooms where experiments are planned and conducted, the place where science plays out and unfolds knowledge and understanding, the researcher and his or her group are defined by more than just what is within the laboratory walls. This concept applies to my own laboratory, the Breast Cancer Research Laboratory (BCRL). It began in 1973 as a small lab consisting of Irma H. Russo and myself. At the time it was a component of the Department of Biology at the Michigan Cancer Foundation in Detroit, and our focus was the study of the pathogenesis of breast cancer using an experimental model. This model was based on Charles Huggins' finding that a single intragastric administration of DMBA (7,12–dimethyl–benz(a)anthracene) induces mammary carcinoma in Sprague–Dawley rats. The BCRL grew slowly as an entity, together with my new responsibilities, first as a Chief of the Experimental Pathology Laboratory, later on as a member and Chairman of the Foundation's Department of Pathology, as well as being a Clinical Associate Professor of Pathology at Wayne State University Medical School, the BCRL kept expanding, not only in personnel and grants support, but by developing its own entity as a research laboratory. Physically, the laboratory was housed at the Michigan Cancer Foundation (now the Karmanos Cancer Center) until 1991. When I was appointed Chairman of the Department of Pathology in the Medical Division at Fox Chase Cancer Center, located in the Northeast section of Philadelphia, almost the entire membership of the original BCRL also made the move to Philadelphia. Even though the BCRL was part of the Pathology Department at Fox Chase, we continued expanding our lines of research until eventually we were established as a separate entity in 1994.

The Laboratory as a Unit of Work with a Defined Leadership and Goal

The laboratory provides the physical tools required for research ideas to flourish, but they are not permanent structures. In most cases, the laboratory only lasts as long as there is leadership and goals shaping and driving it. Research laboratories depend on principal investigators who are able to procure external funding. Because funding depends on political and economical fluctuations, it is not uncommon for laboratories to close. Some make the news, but the majority disappear silently, like ghosts, and nobody remembers they once existed. Often the only evidence of their existence is the annual reports published by their institutions, or the remaining account logs kept by administrative officials. In a 2007 article published in *The Scientist*, Alison McCook[1] notes that more than 4,000 NIH-funded researchers were denied grant renewals, and that for some that meant closing their laboratory doors.

The important concept that the scientific apprentice must remember is that the Laboratory *and* the Researcher(s) are one, and as discussed in the previous section, the research idea or theme, is the driving that makes the laboratory a living entity. In my own laboratory, the BCRL, the theme of our research is what drives us. For example, the seminal studies in breast cancer prevention performed in Michigan which lead us to study the pathogenesis of mammary carcinoma, first in the rodents[2] and then in humans, were related to our interest in developing an *in vitro* system to test the transforming abilities of different genotoxic agents. This has remained an active area of research within our group and grown to the point where we are identifying the genes responsible for the expression of immortalization and transformation phenotypes.[3–16] After moving our lab to Fox Chase Cancer Center, we developed an *in vitro* experimental system to study the transformation of human breast epithelial cells (HBEC) with chemical carcinogens and 17β-estradiol and its metabolites.

I will describe, as an illustrational example, how the studies performed in the BCRL led to the conclusion that in order to induce full neoplastic transformation, (i.e., expression of advantageous growth, anchorage independence, enhanced chemo-invasiveness, absence of ductulogenesis, and tumorigenesis in a heterologous host), the cells need to be immortalized prior to carcinogen exposure. The *in vitro* model of cell transformation was developed utilizing the immortalized human breast epithelial cell (HBEC) MCF-10A, transfected with c-Ha-ras oncogene, and MCF-10F treated with the carcinogen benzo(a)pyrene (BP). The importance of the MCF-10A is that it was derived without viral or chemical intervention from mortal diploid HBECs with an extended life span.

This cell line, now widely utilized worldwide, along with Dr. Fulvio Basolo, a postdoctoral fellow at the time, enabled us in 1990 to determine that the insertion of an activated Ha-ras oncogene alone was capable of inducing malignant transformation in breast epithelium.[4-9] MCF-10A cells were transfected with the c-Ha-ras oncogene contained in the plasmid pHo6-Tras. Their growth pattern on plastic, ability to grow tri-dimensionally in collagen, the expression of anchorage-independent growth, and independence from hormones and growth factors for proliferation, and tumorigenicity in nude mice were then tested and compared with those of MCF-10A parent cells, or cells transfected with the neomycin-resistant gene alone or with the proto-oncogene (pHo6neo). Among the phenotypic markers indicative of *in vitro* cell transformation, the ability of cells to grow in an anchorage-independent manner is a reliable criterion especially when accompanied by tumorigenicity, since many normal cells are able to grow in methocel or agar, but are unable to produce tumors. In MCF-10-neoT cells, however, this property was detected early, after the third passage post-transfection, and was associated with tumorigenicity in nude mice, a phenomenon also expressed by breast tumor cell lines such as MDA-MB-231, BT-20, and MCF-7 cells, other well established cancer cell lines. The relevance of Ras in human breast cancer has been examined and quantitated by analyzing breast tissue samples for expression of ras related mRNA and p21 ras protein, which has been found to be expressed in biopsies of both normal and malignant breast tissues. Whether the ras oncogene is a causative agent of human breast cancer has not been proved, yet the model was extremely useful in generating and answering many different questions on the role of growth factor in cell transformation and neoplasia.[4-9]

We no longer use the c–Ha–ras model but many other researchers still use the basic model to develop *in vivo* models of breast tumorigenesis[19] based on the seminal data created and developed in the BCRL. More than 100 publications have been generated using the ras transformed MCFIA cells. For us the studies performed using c–Ha–ras were a tremendous achievement, not only because they transcended our laboratory, but also because they allowed us to establish the expertise we needed to explore other areas that we are still actively pursuing such as the role of environmental agents or substances like natural estrogens, which have been linked to breast cancer. In 2001 and 2002 we published a seminal paper[17,18] showing that 17β–estradiol, a natural estrogen, was able to transform human breast epithelial cells and further demonstrate that the presence of estrogen receptor alpha was not needed for the transforming event. The model was also useful in showing the mechanism of epithelial mesenchymal transition and the molecular events involved in the process.[19,20]

Based on the experience accumulated with the model of *in vitro* transformation of the human breast epithelial cell MCF-10A with c–Ha–ras oncogene, we decided to explore the important question of whether the chemical carcinogens that induce tumors in animals were able to transform the human breast epithelial cells *in vitro*. These studies were possible because of two perseverant and hard working postdoctoral fellows in the BCRL, Dr. Josiah Oschieng and Dr. Gloria Calaf. With Josiah we were able to study the properties of calcium in the mechanisms of growth and immortalization of MCF10F cells, and with Gloria we were able to transform MCF10F cells with chemical carcinogens. For this purpose we used two carcinogenic compounds that require metabolic activation, DMBA and B[a]P, as well as two direct-acting carcinogens, NMU and MNNG. We found that the acquisition of anchorage independence was manifested at 157 days in culture, and the efficiency varied from carcinogen to carcinogen, the highest being for DMBA and BP. NMU was the least efficient in inducing colony formation and did not originate clones; MNNG originated four colonies from which only one clone, M4, emerged at 248 days post-treatment. M4 exhibited greater colony efficiency (CE) than the parental cells, and generated other subclones that emerged by 446 days in culture. We observed that the apparent refractory nature of human breast epithelial cells to express the phenotypes of neoplastic transformation upon treatment with carcinogens *in vitro* could be overcome by isolating and reseeding the colonies formed in agar-methocel, which are further selected by reseeding in semisolid medium. Whereas this selective pressure *in vitro* may not represent the same forces that are operational *in vivo*, our observations indicate that the emergence of neoplastic phenotypes are a continuum expressed only in those cells that exhibit further growth advantage. The clones derived from DMBA and B[a]P-treated cells, (i.e. D3-1 and BPI-E respectively), exhibited greater invasive and chemotactic capabilities than MCF-10F control cells, and both parameters were similar to those of the tumorigenic T24 cells. The clones BP5, BP7 and BP10 presented greater values for invasion and chemotaxis than MCF-10F control cells, but less than BPI-E, D3-1 and T24 cells. The higher chemoinvasive and chemotactic capacity of BP1-E and D3-1 cells was observed in late passages and correlated with their higher colony efficiency in agar-methocel. The cell line BP1-E expressed tumorigenesis in SCID mice.[15]

These studies not only showed that chemical compounds can transform breast epithelial cells, but also that the immortalization event was required for accelerating this process. In addition, these studies were the basis for furthering our understanding of the interactive effect of chemical carcinogens and oncogenes. Therefore, one way to evaluate the contribution of ras genes in the development of the tumorigenic phenotype was to introduce this gene into suitable acceptor cells. Transfection of the non-tumorigenic cell lines, clones D3-1 and BP1, derived from the carcinogen-treated MCF-10F cell lines established by Gloria Calaf, and the MCF-10F cell line with the c–Ha–ras oncogene not only enhanced colony formation in agar-methocel, as well as invasiveness, but induced tumorigenicity with a short latency period in SCID mice.[21]

This brief historical perspective of the experiments on cell transformation is just one way of illustrating how one particular project was part of a larger vision; its role was to create the adequate tools to carry out the main goal of the breast cancer research laboratory — cancer prevention. The *in vitro* model is now a powerful tool for testing the efficiency of preventive agents against breast cancer by determining if the transformation phenotypes are abrogated or reverted by the substances under investigation. This took several years and it was not until 2008 when work in the BCRL performed by Hilal and Mehmet Kocdor and Sandra Fernandez were able to demonstrate that the hormone hCG that is a preventive agent in the rat model was able to abrogate the neoplastic phenotype *in vitro* of cells transformed with 17*β*-estradiol.[26]

Maintaining a Sense of Individuality and Uniqueness is the Overall Goal of the Laboratory

In Chapter 2 we discussed the research idea as an important driving force in the life of the researcher and in his or her laboratory, and in Fig. 1 we traced a time line of our prevention studies that started in 1975. This line of research has provided a seal of individuality to the BCRL. I have used the BCRL as an example, but it could likely be applied to any other laboratory in the world. In the previous paragraphs I have described the connection of the model of cell transformation and prevention studies. It is particularly important to link those studies in the BCRL with previous studies that were equally important in establishing the individuality and uniqueness of the research experience that has been the BCRL.

Our studies in cell transformation were possible because we started working with the MCF-7 cells developed by H. Soule in 1970 at the Michigan Cancer Foundation. It was the important experiments performed at this institution that paved the way for our laboratory's further development and use of this cell line. Since its initial description, MCF-7 has been verified both biochemically and cytologically. The demonstration of human alpha-lactalbumin and estrogen receptor protein in these cells supported that its origin was from human breast tissue. The MCF-7 cell line contains receptor proteins specific for estrogen, androgen, glucocorticoids and progesterone. Using light microscopy we initially observed that MCF-7

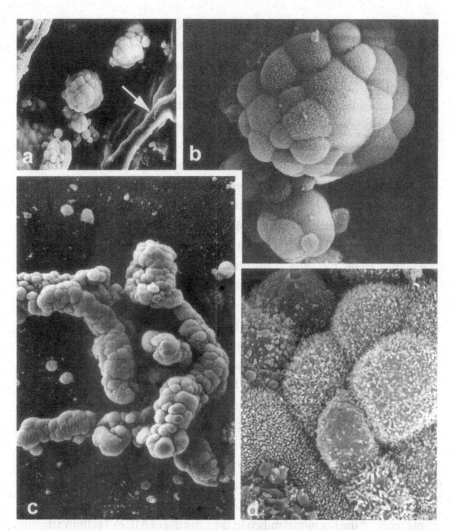

Fig. 1. Scanning electron microscopy of the collagen coated sponge (arrow) and a cluster of cells rising above the monolayer (left), ×400. (b) Larger magnification of the cluster shown in (a), ×3,500. (c) Scanning electron microscopy of cells recovered from the supernatant medium by centrifugation on a Millipore filter showing long arrays of densely packed, small round cells, with sharply defined cell limits, ×2000. (d) Higher magnification of the cell surface of the cells, shown in (c), covered by uniform type of microvilli all over the surface, but microvilli differ widely from cell to cell, ×20,000.

cells were comprised of polyhedral epithelial cells, a finding we confirmed by scanning electron microscopy. Whereas various *in vitro* methods have been used to determine and characterize the malignant potential of tumor cells, MCF-7 cells cultured in a semi-solid media like agar-methocel, form colonies, and when seeded in a collagen matrix they form ball-like structures, or solid masses, of cells indistinguishable from those formed by primary breast cancer cells *in vitro*. A method used by our group was to test the growth properties of breast cancer cells cultivated in collagen-coated cellulose sponges. This provided an excellent technique for the study of three-dimensional expression of tumor morphology, and for investigations into MCF-7's cellular interrelations (Fig. 1). The morphologic pattern exhibited by the MCF-7 cells grown in the collagen-coated cellulose sponge was similar to the histological pattern found in both the antecedent primary tumor and the pleural metastasis from which this cell line was derived.[14] Peter Well and Joseph Saby, both recent college graduates who joined the BCRL around that time, were our first technical assistants. Well and Saby proved invaluable in their help with growing the MCF-7 cells in athymic mice, which resulted in a full demonstration that MCF-7 cells were tumorigenic. Athymic mice are a special strain of BALBC mice born without the thymus, making them immunosuppressed, and therefore able to tolerate exogenous tissues like neoplastic cells. Keeping that particular type of animal alive in those days was not an easy task. It was Joseph and Peter's ingenuity and perseverance in designing and constructing a special rack that maintained the cages in a sterile environment that kept the mice alive long enough to finish our experiments. It was their sense of purpose that made demonstrating the importance of the hormonal milieu in the growing of MCF-7 cells in athymic mice possible.[22] To illustrate the importance of this finding, a criterion for malignancy was the ability of transformed cells to grow in an adequate heterotransplantation system. Inmunologically depressed athymic mice (nu/nu) have the striking capability of discriminating between normal and neoplastic cells. Normal cells do not induce tumors, whereas malignant cells do. Our first experiment demonstrated that MCF-7 cells were unable to grow in male or female athymic mice. The gross examination of the area of cell inoculation and the histological study revealed a complete absence of the inoculated cells;

disorganization of the fat and some fibrosis were the only changes observed. Only those mice that received a transplant of pituitary glands or ovaries from *syngenic* mice induced the growth of MCF-7 cells. With Peter and Joseph, we carefully obtained the pituitaries and ovaries of the syngenic mice and transplanted them in the renal capsule in order to boost the hormonal milieu of the host. Nine of the 11 (82%) inoculated female mice that received pituitary grafts developed palpable tumors within 12 to 18 days after inoculation. The tumors adhered to the skin and underlying muscle. Eight of the 13 (61.5%) inoculated female mice that received ovary grafts developed palpable tumors within 12 to 18 days. The tumors were small, oblong masses of 1.5–2.5 mm at their largest diameter. They adhered to the dermis of the skin and to the muscle of the abdominal wall. The tumors were firm, of a rubbery consistency, and presented resistance to sectioning. The tumor's vascular bed was well developed. The area of the tumor was easily distinguished from the scar produced by the cauterization and the incision made during the transplant procedure. The histological patterns of the 17 tumors studied were identical, while the tumors observed in mice isografted with pituitary glands or ovaries were indistinguishable.[22,23] Later on, a hormonal pellet replaced the ovary and pituitary grafts.[24] We found that the use of an estrogen supplemented castrated male was also suitable for growing MCF-7 cells. The removal of the uterus and supplementation with estradiol, either as pellets or as a silastic tube containing 5 mg of estradiol in female mice is also a standard procedure. The removal of the uterus prevents swelling and accumulation of fluid in the uterus due to the estrogenic stimulation.[23] This seminal experiment established the basis for this model which is used to test antineoplastic drugs in heterotransplanted mice with MCF-7 cells, and has been pivotal to our transformation assay using human breast epithelial cells.

The Laboratory as a Source of Fulfillment

The physical space of the BCRL, as well as its members, has changed with the years, but the desire to blaze a new path has been alive and well since the BCRL's inception in 1973. Since then we have grown into 12 major research projects and have generated more than 350 publications and five books.

Is there a special recipe for running a lab like the BCRL? Yes. The most important component of our research endeavor has been sharing our enthusiasm and desire for the knowledge gained from experimentation with our scientific apprentices. From the more than 50 trainees that have been involved with the lab through the years we know that those who chose scientific research have kept alive the concept of scientific apprenticeship by continuing to train other researchers. This is a major piece of what makes a research laboratory successful. If I were to count the disciples taken on by my former scientific apprentices the number would soar into the triple digits. I am happy to know that so many people have been touched by our work. Is this unique to the BCRL? Almost all of the world's legendary laboratories[25] have provided the intellectual milieu necessary to foster scientific apprentices, and in the end, this is one of the research laboratory's essential missions.

References

1. McCook A. Losing your lab. *The Scientist*, p. 32, May 2008.
2. Russo J, Saby J, Isenberg W and Russo IH. Pathogenesis of mammary carcinoma induced in rats by 7,12-dimethylbenz(a)anthracene. *J Natl Cancer Inst* 59:435–445, 1977.
3 Russo J, Reina D, Frederick J and Russo IH. Expression of phenotypical changes by human breast epithelial cells treated with carcinogens *in vitro*. *Cancer Research* 48:2837–2857, 1988.
4. Basolo F, Elliott J, Tait L, Chen XQ, Maloney T, Russo IH, Pauley R, Momiki S, Caamano J, Klein-Szanto AJP, Koszalka M and Russo J. Transformation of human breast epithelial cells by c-Ha-ras oncogene. *Molecular Carcinogenesis* 4:25–35, 1991.
5. Basolo F, Elliott J and Russo J. Transformation of human breast epithelial cells with foreign DNA using different transfecting techniques. *Tumori* 76:455–460, 1990.
6. Ochieng J, Basolo F, Albini A, Melchiori A, Watanabe J, Elliott JW, Raz A, Parodi S and Russo J. Increased invasive chemotactic and locomotive abilities of c-Ha-ras transformed human breast epithelial cells. *Invasion and Metastases* 261:1–10, 1990.

7. Ciardello F, McGeady ML, Kim N, Basolo F, Hynes N, Langton BC, Yokozaki H, Saeki T, Elliott J, Mauri H, Mendelsohn J, Soule H, Russo J and Salomon D. Transforming growth factor alpha expression is enhanced in human mammary epithelial cells transformed by an activated c-Ha-ras protooncogene but not by the c-neu protooncogene, and overexpression of the transforming growth factor alpha complementary DNA leads to transformation. *Cell Growth Differ* 1:407–420, 1990.

8. Rak J, Basolo F, Elliott J, Russo J and Miller FR. Cell surface glycosylation changes accompanying immortalization and c-Ha-ras transformation of normal human mammary epithelial cells. *Cancer Letter* 57:27–36, 1991.

9. Maloney T, Geronimo I, Fontanini G, Basolo F, Elliott JW and Russo J. Transfection of a normal human breast epithelial cell line with c-Ha-ras oncogene alters the expression of p21 and cytokeratins. *Am J Pathol* 140:1483–1488, 1992.

10. Russo J, Tait L and Russo IH. Morphologic expression of cell transformation induced by c-Ha-ras oncogene in human breast epithelial cells. *J Cell Sci* 99:453–463, 1991.

11. Ochieng J, Tahin QS, Booth CC and Russo J. Buffering of intracellular calcium in response to increased extracellular levels in mortal, immortal and transformed human breast epithelial cells. *J Cellular Biochem* 46:1–5, 1991.

12. Basolo F, Serra C, Ciardello F, Fiore L, Russo J, Campani D, Dolei A, Squartini F and Tomiolo A. Regulation of surface differentiation molecules by epidermal growth factor and transforming growth factor a in human mammary epithelial cells transformed by an activated c-Ha-ras protooncogene. *Int J Cancer* 52:1–7, 1992.

13. Russo J, Calaf G and Russo IH. A critical approach to the malignant transformation of human breast epithelial cells. *Crit Rev Oncog* 4:403–417, 1993.

14. Calaf G, Tahin Q, Alvarado ME, Estrada S, Cox T and Russo J. Hormone receptors and cathepsin D levels in human breast epithelial cells transformed by chemical carcinogens. *Breast Cancer Res Treat* 29:169–177, 1993.

15. Calaf G and Russo J. Transformation of human breast epithelial cells by chemical carcinogens. *Carcinogenesis* 14:483–492, 1993.

16. Zhang PL, Calaf G and Russo J. Allele loss and point mutation in codons 12 and 61 of the c-Ha-ras oncogene in carcinogen-transformed human breast epithelial cells. *Molecular Carcinogenesis* 9:46–56, 1994.

17. Russo J, Hu YF, Tahin Q, Mihaila D, Slater C, Lareef MH and Russo IH. Carcinogenicity of estrogens in human breast epithelial cells. *Acta Pathol Microbiol Immunol Scand (APMIS)* 109:39–52, 2001.

18. Russo J, Lareef MH, Tahin Q, Hu Y-F, Slater CM, Ao X and Russo IH. 17-beta estradiol is carcinogenic in human breast epithelial cells. *Steroid Biochem Molec Biol* 80(2):149–162, 2002.

19. Russo J, Fernandez SV, Russo PA, Fernbaugh R, Sheriff FS, Lareef HM, Garber J and Russo IH. 17-beta estradiol induces transformation and tumorigenesis in human breast epithelial cells. *FASEB J* 20:1622–1634, 2006.

20. Huang Y, Fernandez S, Goodwin S, Russo PA, Russo IH, Sutter T and Russo J. Epithelial to mesenchymal transition in human breast epithelial cells transformed by 17-beta–estradiol. *Cancer Res* 67:11147–11157, 2007.

21. Calaf G, Zhang PL, Alvarado MV, Estrada S and Russo J. c-Ha-ras enhances the neoplastic transformation of human breast epithelial cells treated with chemical carcinogens. *Int J Oncol* 6:5–11, 1995.

22. Russo J, McGrath CM, Russo IH and Rich MA. Tumoral growth of a human breast cancer cell line (MCF-7) in athymic mice. In *III Int. Symp. on Detection and Prevention of Cancer*, Nieburgs HE (Ed.). New York, pp. 617–626, 1976.

23. Russo J and Russo IH. *Biological and Molecular Basis of Breast Cancer*. Springer-Verlag, Heidelberg, Germany, 2004.

24. Shafie SM and Giartham FH. Role of hormones in the growth and regression of human breast cancer cells (MCF-7) transplanted into athymic mice. *J Natl Cancer Inst* 67:51–56, 1981.

25. Burke A. Legendary Labs. *The New York Academy of Sciences Magazine*, Winter 2008.

26. Kocdor H, Kocdor MA, Russo J, Snider KE, Vanegas JE, Russo IH and Fernandez SV. Human chorionic gonadotropin (hCG) prevents the transformed phenotypes induced by 17 β-estradiol in human breast epithelial cells. *Cell Biol Int* 33:1135–1143, 2009.

Suggested Readings

Ault A. Postdocs tangled up in red tape. *The Scientist*, p. 42, February 14, 2005.

Best place to work 2008 post docs. *The Scientist*, 53, February 14, 2005.

Gallagher R. An iGEM of an idea. *The Scientist*, p. 13, December 2007.

Nelson S. The Harvard computers. *Nature* 455:367–337, September 4, 2008.

Nurse P. US biomedical research under siege. *Cell*, pp. 9–12, January 13, 2006.

One woman is still not enough. *Nature* **451**: 865, February 21, 2008.

Rothblatt S. The professor comes of age. *American Scientist* 94:466, September–October 2006.

Shaping life in the lab. *Times*, p. 865, March 9, 1981.

The best places to work 2007 academia. *The Scientist*, p. 61, November 2007.

The best places to work 2008 academia. *The Scientist*, p. 47, November 2008.

The scientific balance of power. *Nature* **439**:646–647, February 2006.

Wiesel T. Balancing biomedicine's postdoc exchange rate. *Science* **289**:867, August 11, 2000.

Interactive Groups in Scientific Research

The apprentice of science can learn research through both the *individual exchange* and *interactive groupings*. These arbitrary classifications are based on the way that the scientific apprentice interacts with the mentor, or principal investigator (PI), of the laboratory. In the *individual exchange*, the interaction usually occurs directly between the apprentice and the mentor, a quality that uniquely defines this kind of relationship. *Interactive groupings*, however, are defined by direct and indirect communication between multiple people.

The Individual Exchange

In this kind of relationship the scientific apprentice has direct contact with the mentor, or PI, who serves as their main source of input. This does not mean that the PI is the only one who influences the scientific apprentice, rather the main one, the effect being a deepened rapport between the two. This relationship is centered on the specific targeted area of research in which the scientific apprentice has initiated his or her work, or the initiation of a unique idea developed in partnership with the PI. This defined realm provides the scientific apprentice exclusive ownership over an idea that, ultimately, will foster a secure publication of which in most cases, the scientific apprentice will be named first author. Whereas this kind of interactive grouping does not preclude that the PI have other similar one-on-one interactions with other scientific apprentices, it is unlikely that a scientific apprentice will interact in the same meaningful way with anyone else.

How this special relationship emerges between the scientific apprentice and their mentor is difficult to generalize, and even more difficult, if not impossible, to standardize into a formula that can be applied to everybody. I was fortunate that my first mentor was generous enough to teach me what he knew and help me learn the basic steps that proved to be so fundamental later on in my career.

The year of 1961 marked my second year of medical school and I was determined to be a medical researcher. Fortunately, two opportunities came to me that very same year. Dr. Mario H. Burgos, Director of the Institute of Histology and Embriology that now holds his name at the University National of Cuyo in Mendoza, Argentina, was my professor of Histology and Embryology. He had been trained by Bernard Houssay and came to the Medical School after a traineeship in electron microscopy with Dawn Fawcett at Harvard University. I was fascinated by the structure of the cell and the modern concepts of cell biology that were available in those days. The double helix was still the big discovery of the century, so were the new findings of cell organelles and their function, such as the role of the rough endoplasmic reticulum in the synthesis of protein, the world of the mitochondria and the lysosomes. These discoveries were to provide the basis for the functions and the localization of molecules in every part of the cell. I already knew *The Book of Cell Biology* by Eduardo De Robertis[1] by heart since I had been reading it with fascination since my fourth year of high school when I was studying for my preparatory examination to enter medical school. Burgos was an excellent teacher and had an extraordinary talent for drawing and painting that he transmitted in his classes. I think I must have somewhat impressed him too with my enthusiasm — I was ready to ask questions that had arisen in my previous readings of *Cell Biology*. I still remember those days with great fondness. The wonderful world of science was opening itself to me completely, and showed a beauty far more brilliant than I could have ever expected.

Professor Burgos asked me to join his group that year. The Institute, short for Institute for Histology and Embryology, was a nest of young scientists, working in various areas of cell and structural biology, and each one enthusiastically tackled their project with the fervent energy transmitted by

Burgos. Around that same time I met Dr. Julian Echave Llanos, a Professor of Pathology and Director of the Experimental Pathology Institute at the same medical school. I met Echave Llanos through a dear friend, Ramon Piezzi, who was also hooked on research and happened to be one of his students. To my surprise, Echave Llanos remembered me; he had been the one who evaluated my admission test into medical school. Perhaps he sensed, even then, the fire that was in me. It wasn't until years later that Echave Llanos confided to me that my person oozed dedication and an obsessive passion for research in those days.

I lay awake many nights pondering whom to choose as a mentor. The trouble was, while I respected and admired Burgos, the Experimental Pathology Institute was where the process of diseases such as cancer was studied, and this, more than anything else is what fascinated me. Despite the beauty of histology and embryology, I said no to Burgos' invitation, although I kept a good connection with him and the researchers in his group. In retrospect it was a good decision. Echave Llanos had recently returned from a sabbatical in Germany, bringing the newest tools available in experimental pathology in those days to the medical school. He was an extraordinary mentor and teacher for me. I was the youngest of the group and he was the one who taught me how to do research, from the simple task of handling a mouse for injection, to performing surgical procedures, planning an experiment and obtaining conclusions. He taught me the value of statistical power, reproducibility and quality control. He taught me how to construct scientific tables and figures, and how to make graphics that clearly express data. I absorbed his influence like a thirsty sponge. To this day I wonder whether he ever realized how much I learned from him and how deeply I have appreciated his mentorship.

This is just one example of an *individual exchange*, which I maintained for many years of my scientific apprenticeship. It was during this interaction that I authored my first four scientific publications, and where I learned how to present and discuss a paper in a scientific meeting. This interaction was extremely valuable to me and it is in the scientific apprentice's best interest to find a mentor whom they can learn from and work closely with. One of the most important aspects of this formative period for me was that very early on I learned how to handle complex biological processes, one of them being tissue regeneration, liver regeneration in particular. Of specific interest in this area was whether a physiologic process, such as circadian rhythm, controlled the process of liver regeneration. One of my first contributions was demonstrating that the process of liver regeneration was well controlled by light and darkness during the first 48 hours after a partial hepatectomy. The results of this work were published in *Zietschrift fur Zellforschung.*[2] The study of circadian rhythm lost momentum, however, and it wasn't until relatively recently, with the discovery of clock genes, that the scientific community started to take a second look at this process. In 2003, I acquired funding to study the circadian rhythm of clock genes in the mammary gland of the rat and mentored Dr. Johana E. Vanegas, a scientific apprentice in my laboratory with this project.

Individual exchanges, however, must be limited, otherwise they could be detrimental to the scientific apprentice in the sense that he or she may become too dependent on the mentor, ultimately narrowing his or her long-range intellectual growth. In my case it was after five years at the Institute of Experimental Pathology that I started to realize that it was time for me to generate my own line of research. I wanted to loosen the restrictions of the mentorship and initiate my own ideas. Echave Llanos on the other hand wanted to maintain an *individual exchange* type of relationship with all his disciples which left little room for us to interact and collaborate with each other, not to mention any other groups. At that point I was at the end of my medical courses and needed a serious plan of action. The first thing to do was to get my degree as a Medical Doctor, an optional degree and an addition to the Physician and Surgeon degree granted by the medical school. A doctoral thesis based on an original and independent scientific work was required for this, however. The second task was to receive a fellowship from the National Council of Research in Argentina, the only institution that supported those who wanted to do scientific research at the time. With these objectives in mind I paid Professor Burgos a visit and soon after moved to his institute. I maintained an *individual exchange* type of interaction with Burgos, and he was my mentor for the following four years. Burgos was a different type of mentor than Echave Llanos; he outlined general ideas and let people work from that point forward. He was there when we needed him, but he allowed us the intellectual freedom of our own research interest and that was exactly what I was looking for.

My own style of mentorship has in the past followed the *individual exchange* model based on my experience with Echave Llanos. Lately my schedule does not allow me the luxury of spending so much time one-on-one with each post-doc but in order to maintain a close mentor–apprentice relationship, I have adopted Burgos' style. I entrust, however, most technical aspects of the work to the associates who I feel would approach the transmission of knowledge and skill in the same way I would. If it were at all possible I would take the time to work individually with each person, yet by giving my scientific apprentices room to think and grow, I think I have given them something much more valuable.

By the time I left Argentina in 1971, The Burgos Institute was composed of physician scientists all at different levels of their research endeavor. Some were already independent investigators, others were associates (meaning they were under supervision of a more senior or independent investigator) and others were scientific apprentices like me. Because of the diversity of experiences and themes it was easy to learn from all of them by listening and discussing their triumphs and mistakes. It was in this period that I consolidated my training in cell biology and learned all the ancillary techniques available in those days, including electron microscopy, tissue culture and developmental biology and embryology. The latter subject was of special interest to me because if was through this discipline that I met Dr. Irma H. Alvarez, who would eventually become my wife and partner in science. It is because I had such an enriching experience under Echave Llanos' and Burgos' mentorship that I have tried to follow their example with the 54 trainees I have mentored through the years.

Before we leave our discussion of *individual exchanges* there are a few things to consider before committing oneself to this kind of relationship. It is important that the scientific apprentice critically evaluate the surrounding intellectual milieu, such as the progress of other scientific apprentices, their publications and work, as well as how they master their fellowship and further learning. This says a lot about the mentor as well as gives a sense of the kind of environment he or she creates around his or her work. One should also try to determine the degree of initiative that other scientific apprentices have in the group and how they manage their data and intellectual growth because you will inevitably be learning from them as well.

The Interactive Groupings

In this kind of working relationship the scientific apprentice still has direct contact with their PI, or mentor, but they are also part of a team and are actively involved in working towards the solution to a specific scientific problem. The importance of this kind of group interaction lies in the fact that it can incorporate multiple areas of research and disciplines. There is an obvious advantage to working towards a common goal: expertise is broadened and ideas develop more quickly. But these efforts also require flexibility, team spirit, and based on the complexity of the task, more combined working hours. By being a part of this interactive group the main advantage is that each member of the team secures not only co-authorship in the publication, but most likely several co-authorships. However this type of interaction is not suited for everybody. It is the wisdom of the mentor to identify those who are team players suited *interactive groupings* from those that are more interested to maintaining an *individual exchange*.

The Scientific Apprentice as Part of the Research Team

Whether the scientific apprentice is part of a large or small team, the theme established by the PI should be in the scientific apprentice's same area of interest. The scientific apprentice becomes a member of the team and channels his or her ingenuity and strength towards the service of the established objective. The Manhattan project might serve as a model for *interactive grouping*, in that the development of the atomic bomb was an undertaking in which great expertise and independent minds where put to work towards a monumental, groundbreaking goal. The collaboration required for this project has been imitated over and over again, but the outcome is not always as successful (or as contentious). We can considerably scale down this model to discuss the essence of *interactive grouping* which every scientific apprentice can be part of.

As long as one of the aims of the team or the group match with the scientific apprentice's area of interest, their scientific endeavor can be successful. Based on this premise I encourage any scientific apprentices to follow this route early in their carriers because often times it serves as an important learning experience. It is rewarding to be part of a good team and learn from more senior investigators, not just the mentor, the skills it takes to make science happen.

I went to the Michigan Cancer Foundation (MCF) in Detroit in August of 1973, after a two-year fellowship granted by the Rockefeller Foundation, to be part of a research team. I had been invited to join the staff of scientists working on a project in which my expertise was needed. The goal of the project was to search MCF7 cells at the ultra-structural level for the presence of a retrovirus. MCF-7 is a breast cancer cell line isolated in 1970 from a 69-year-old Caucasian woman. Dr. Michael Brennan, the president of the institution at the time, was treating the patient, who also happened to be a nun. She had what appeared to be a tumor mass in her right breast and the tissue removed was not confirmatory of cancer. Five years later, however, after noticing a lump/discoloration, a second operation revealed a malignant adenocarcinoma. Over the next three years local recurrences were treated with radiotherapy and hormonotherapy. In the process of removing the chest wall nodules, a pleural effusion was discovered. Herbert Soule, a staff scientist at the Michigan Cancer Foundation, plated the cells of the pleural effusion and established the MCF-7 cell line, the number seven marking Herbert Soule's seventh attempt at generating a breast cancer cell line. This cell line retained several characteristics of differentiated mammary epithelium, including the ability to process estradiol via estrogen receptors, and the capability of forming domes. In addition, when cultured in a tri-dimentional matrix (3D), the cells reproduced the same cluster of neoplastic cells as those found in the pleural effusion (see Fig. 1 in Chapter 5).

Searching for the retrovirus in the MCF7 cells required the use of an electron microscope, a tool that enabled us for the first time to see the internal structure of the cell. I was quite familiar with the electron microscope because of my training in ultra-structural cell biology with Burgos. The quest for the retrovirus was part of larger project funded by the National Cancer Institute. The thinking was that if a virus could be identified in this established human breast epithelial tumor cell line, it would link to the viral etiology of breast cancer. Bittner in 1930 had already determined that mammary tumors in certain strains of mice were caused by a virus, what we know as the Bittner Factor. The idea was that the identification of the retrovirus in the MCF-7 cells would confirm that the Bittner factor was a causative agent of breast cancer.

Excitement about the project also came from Howard Martin Temin, Renato Dulbecco and David Baltimore's seminal discovery of reverse transcriptase for which they shared Nobel Prize in Physiology or Medicine in 1975. Reverse transcriptase explained how tumor viruses act on the genetic material of the cell and introduced a revolutionary concept that contradicted the "Central Dogma" of molecular biology that genetic information flows exclusively from DNA to RNA to protein. Certain tumor viruses carry the enzymatic ability to reverse the flow of information from RNA back to DNA, using reverse transcriptase. Reverse transcriptase is the central enzyme in several widespread human diseases, such as HIV, the virus that causes AIDS.

From 1973 to 1976 I was immersed in the world of big science, and saw how scientists like Gallo, Rausher, Spiegelman, Mulkbrock, Scarpelli, and others, were presenting evidence and new possibilities in favor of and against the viral hypothesis. Another great moment in science came with the signing of the 1974 Cancer Act, which significantly increased the NCI's appropriations, and what had only once been a dream, due to financial constraints, was made reality and the sky seemed the limit.

Ultimately our studies were unable to isolate the virus or to demonstrate a causal relationship between the retrovirus and cancer, nor were we able to develop a vaccine. There are still researchers working in this area who sustain that a large proportion of human breast cancers may be associated with the human mammary tumor virus (HMTV). This human virus is nearly identical to the murine mammary tumor virus (MMTV) that is implicated in breast cancer in mice.

My contribution to virology was very small, but I am grateful to my colleagues at the time, exemplary scientists such as Marvin Rich, Michael Brennan, Charles McGrath, Philip Furmanski, Herbert Soule and Justine McCormick. All of them shared so much of their experience with me; they cannot imagine how much of what I have learned has come from them. To be surrounded by and collaborate with scientists that act and think differently is a wonderful experience for every scientific apprentice — it certainly was for me. Even though my research idea did not directly stem from my work with them, it was in living the experience that I found my own way. As I described in Chapter 2, it was while immersed in this milieu that I found my own research interests, ideas that continue to challenge and exhilarate me to this very day.

Dynamic of the Research Team and Grantmanship in Interactive Groupings

Original discovery. Sometimes it is an isolated event, like Alexander Fleming's more or less accidental discovery of the *Penicillium notatum* that was synthesized into an active product later called penicillin. Perhaps due to an inadvisable carelessness, Fleming was prone to accidental discoveries such as the one that brought us the first antibiotic. In 1922, while working with some bacteria, a drop of mucus leaked from his nose. When he returned a few hours later, the bacteria had disappeared. He had discovered a natural substance found in tears and nasal mucus that helps the body fight germs — the lysozome. Although others had, independently of Fleming, pursued studies of the lysozyme, he made the important discovery that a substance could kill bacteria without affecting the human body. Alexander Fleming was the one who made many important observations, but he needed the entrepreneurial aspect of others to acquire the funds to make his discovery relevant. Today we look at that entrepreneurship in terms of having the ability to get a grant or the resources to make it happen (grantmanship). From the point of the initial revelation however, other players, equally important, will need to turn that original observation into funding in order to follow through with testing and confirming the hypothesis. In penicillin's case it was the vision of Howard Florey, the collaboration of Ernst Chain, and Norman Heatley's microbiological craft that propelled the marketing and industrialization of this powerful weapon against infectious diseases. All these minds were critical and without them so many lives would not have been saved.

How Successful are Interactive Groups?

John Whitfield in his 2008 article "Group Theory"[3] pointed out that interactive groups can be considered successful when measured by their number of publications in *Nature*, a journal where only six single-author papers, out of a total of some 700, were identified in a single year. According to Whitfield, the proportions are similar in any other leading research journal. The reason for the surge in collaborative science is in a certain way driven by the complexity of the technologies that advance at such a fast pace that it is difficult for one laboratory to maintain competitive expertise. This trend is also driven by the fact that granting institutions like the NIH, the Department of Defense or The Komen Foundation (to cite just a few) foster these kinds of relationships.

But perhaps there are other reasons for collaboration that stem from the individual scientists and are not driven only by peer pressure, technology and granting feasibility. Through the years I have established collaborative links that have produced rewarding experiences in obtaining funding and solving difficult problems as well. The movement towards this style of working is in a sense a response to the increased complexity of the science itself. Our predecessors lived in a different scientific environment, one in which the focused work of a single scientist working in isolation could change society's entire understanding of nature. As modern scientists our work is much more complex in that we have a real knowledge of biological processes and there are more of us in the world communicating at a lightning fast rate. The only way to grow in this field is to talk to our peers and aim to solve problems by moving out of our cocoons and seeking out collaborators.

I first saw vivid evidence of this new way of working on April 22, 1996 in a small meeting room of The Renaissance Hotel in Washington DC. Eighteen researchers* from different disciplines and academic institutions were brought together for the first time by David Longfellow of the National Cancer Institute, Ercole Cavalieri from the University of Nebraska and Joachim Liehr from The University of Texas at Galveston. The objective of the meeting was to understand how estrogens act as endogenous carcinogenic agents in the breast. The idea was to form a Complementary Collaborative Coalition that from then on was called the Cancer Cube. The Cancer Cube did not provide funds, but its mission was to address the important questions that would lead us to the right scientific approach to study how estrogen induces carcinogenesis in the human breast. We met twice a year, and year after year we were clarifying in our minds the complexity of estrogen's role in breast cancer. We finally succeeded in putting our ideas together and five of us, Cavalieri, Liehr, Santen, Sutter, and myself presented a grant application to the Department of Defense. We were granted sufficient funds to form a Collaborative Breast Cancer Center that lasted for five years and generated several publications and collaborative links that are maintained to this day. Working in such a heterogeneous group was not an easy task and had its difficult moments. For people from such different disciplines to see the validity of other's viewpoints can be trying. It became clear in that first phase of the interactive group that there were three important ingredients: congeniality, respect for the knowledge of the others, and a genuine interest in learning from others across discipline and fields. Those were the most valuable lessons I learned and they allowed me to form new collaborative groups around the world, most of which have been successful in obtaining significant grants that would otherwise be out of reach to an individual investigator.

---

* Maarteen Bosland, Ercole Cavalieri, Bob Crevelin, Krystyna Frenkel, Hank Gardner, Shuk-Mei Ho, Joachim Liehr, Leo Liu, Don Malins, Elli Rogan, Deoduta Roy, Tom Sutter, Juthit Weisz, Jim Yager, David Longfellow, myself, and later on Colin Jefcoate, Shutsung Liao, Richard Santen and Richard Weinshiboum.

References

1. De Robertis EDP, Nowinski WW and Saez FA. *Citologia General, Libreria el Ateneo*, 3rd ed., 1955.
2. Russo J and Echave-Llanos JM. Twenty-four hour rhythm in the mitotic activity and the water and dry matter content of regenerating liver. *Zietschrift fur Zellforschung*(2) 161:824–828, 1964.
3 Whitfield J. Group theory. *Nature* 455:720–723, October 8, 2008.

Suggested Readings

Baltimore D. A global perspective on science and technology. *Science* 322:544–511, October 24, 2008.

Branscomb LM. Research alone is not enough. *Science* 321:915–916, August 15, 2008.

Community cleverness required. *Nature* 455:1, September 4, 2008.

Dealing with democracy. *Nature* 425, September 25, 2003.

Endersby J. Command and control. *Nature* 453:721–722, June 2008.

Enserink M. Will French science swallow Zerhouni's strong medicine? *Science* 322:1312, November 28, 2008.

Finn O. Directing a life of science. *Science* 321:906–907, August 15, 2008.

Henney A and Superti-Furga G. A network solution. *Nature* 455:730–731, October 9, 2008.

Mervis J. Science scholarships go begging. *Science* 321:908, August 15, 2008.

Newfield C. Saving public universities. *Nature* 455:467–468, September 25, 2008.

Nuzzo R. Seeing the pattern. CR, pp. 51–53, Spring 2008.

Stengel R. Mandela, his 8 lessons in leadership. *Time Magazine*, July 21, 2008.

Team assembly mechanisms determine collaboration network structure and team performance. *Science* 308:697–702, April 29, 2005.

Wells J and Woodley M. A populist movement for health. *Science*, Vol. 32, October 3, 2008.

Chapter 7

Laboratory Protocol Books

Defining a Protocol Book

The laboratory protocol book is where the documentation of work performed on daily basis is recorded, the primary source for reports and publications and place for reflecting on the research endeavor. Beginning on the scientific apprentice's first day in the research laboratory, taking notes becomes a daily process. The skill of note-taking develops over the years and extends naturally into a young researcher's daily life. With the lab notebook, or protocol book, being such a vital piece of lab equipment, one might think it is redundant to discuss the protocol book any further. Yet the protocol book is vital, not only as a place to write things down, but as a document.

In thinking how to go about organizing the protocol book we must first think about how to organize our research work. For the biologist, and in that I include all biomedical scientists, our work is divided into experiments, or protocols. An experiment is a very simple procedure performed in the laboratory and includes anything from the simple descriptive work of histological procedures, to the most sophisticated experiment in molecular DNA recombination.

In its most basic sense the term "protocol" refers to a detailed plan of a scientific or medical procedure. In order to describe how to stain tissue sections and ensure consistent results each time the procedure is performed, one needs to write a protocol detailing each of the steps. A well-written protocol is crucial in that most often the researcher carrying out the experiment or procedure must rely solely on the text. Ambiguity is a recipe for erratic results. Moving deeper into another dimension of the term's meaning, we see that a protocol also refers to instances in which human involvement, such as tissue collection or clinical intervention, is central to the experiment. In this case we call it a *clinical protocol*.

When designing a project the first step is to provide the experiments, or protocols, a sequential number. Numbering experiments is the most convenient way to differentiate between them and organize the material generated under that experiment, as I will describe later on in this chapter. At the present time in our laboratory we are on experiment Number 891; experiment number one was conducted in 1973 and dealt with the search for a viral article in MCF7 cells. By numbering each one sequentially we have been able to easily keep track of a constantly expanding body of work and able to trace the way each one relates to the other. The numbering of experiments has significant importance when the number of samples collected reflects different treatments and procedures.

I advise beginning a *logbook of experiments*, a place to record experiment numbers followed by brief titles describing them, the date of initiation, and the name of the investigator(s) involved. A logbook helps avoid the confusion that can arise if an experiment is assigned two different numbers or the same number to two different experiments. The Laboratory Director, or the person that oversees the division of work in the laboratory must control the logbook. If the logbook is generated with the assistance of a computer system, a back-up system is a must.

While the protocol book as I describe it here is not adequate for a large toxicological study, like the studies performed by the National Toxicology Program, for example, it is however a must for most independent investigators, starting with the scientific apprentice. The first thing you need to start a protocol book is a laboratory notebook. An adequate lab book includes numbered pages with carbon sheets (such as laboratory research notebook [43-649] distributed Dennison Stationery Product Co., Framingham, MA 01701). The first three full pages of the notebook should be used for the index. Identify the owner of the book and date of initiation and termination.

The Simplest Experimental Protocol

Below is an example of an actual experimental protocol performed in our laboratory that contains the basic information needed: experiment number, date of initiation, investigators involved, objective, material and methods, and expectations of the report. This is the kind of experimental protocol that will provide initial data on a testing of a compound, in this case, *Chloroquine*. It is also called a pilot project intended to provide a sense of what direction additional projects will go. Most of these pilot experiments give a very good idea whether it is wise to pursue that particular line of bioresearch or not. Writing an experimental protocol can serve as an important training tool for the scientific apprentice.

Protocol Experiment 848
Date: March 14, 2008
Investigators:

1. Objective:

To determine if the anti-malarial agent Chloroquine has a preventive effect in DMBA mammary carcinogenesis.

2. Material and Methods
2.a. Experimental Schedule

Female Sprague Dawley rats at 50–55 days of age will be separated in three groups of 20 animals each:

Group 1 will be inoculated with Chloroquine (CQ)(Sigma, St Louis), 45 mg/kg i.p for 21 days daily (Fig. 1).

Group 2 will be inoculated with Chloroquine (CQ) (Sigma, St Louis), 45 mg/kg i.p for 21 days daily. And at the end of treatment (Fig. 1), they will receive I.G instillation of 100 mg/kg b.w. of 7,12-dimethyl-benz(a)anthracene (DMBA), in corn oil.

Group 3 will be inoculated with saline (S) for 21 days daily. And at the end of treatment (Fig. 1), they will receive I.G instillation of 100 mg/kg b.w of 7,12-dimethyl-benz(a)anthracene (DMBA) in corn oil.

2.b. Sample Collection

2.b.i. Mammary glands: At the end of 21 days of treatment with Chloroquine or saline, two (2) hours previous to euthanasia, five (5) animals per group will receive BrdU, 100 mg/kg b.w. i.p. for the study of cell proliferation. At the euthanasia the mammary glands will be collected for whole mount, histology and determination of the labeling index using the BrdU.

2.b.ii. Tumorigenesis: The animals, 15 for each group, will be followed weekly for tumor appearance.

2.b.iii. Determination of tumor response: The three groups of animals will be followed for six months. At the end of this time all the animals will be euthanized and the normal mammary glands and tumors will be dissected, fixed in 10% formalin buffer and prepared for histological examination.

The next section of the protocol, the *Report*, is where the type of data the experiment is expected to yield is described. One type of data in this example is the effect of Chloroquine, acting as a differentiating agent, on the mammary gland. This effect will be the first line of evidence that Chloroquine could be used as a preventive agent. The second line of evidence would be whether treatment with Chloroquine reduces the tumorigenic response.

3. Report

Two sets of data will be obtained:

1. A short term effect of Chloroquine in mammary gland morphology and labeling index to asses the direct effect on gland differentiation. This parameter will be measured at the end of treatment (Fig. 1).

2. The second set of data is in tumor response. The total number of animals with tumors, the total number of tumor per animals will be expressed and tabulated together with tumor size. In addition we will evaluate the presence and number of pre-neoplastic lesions, such as ductal carcinoma *in situ* and tumor type. This parameter will be evaluated in the histological sections.

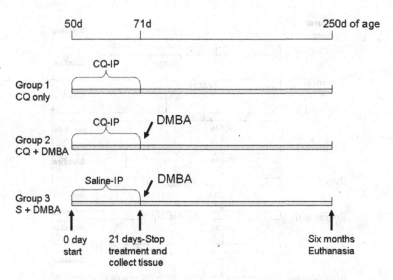

Fig. 1. Schematic representation of a simple experimental protocol.

Example of a Complex Experimental Protocol

In contrast to the previous example, a more complex protocol involving hundreds of animals with different treatment regimens requires careful preparation. The success of it depends on meticulous attention to all the different steps, even those that are considered minor or irrelevant. The example below is a protocol that was used for one of our experiments. As in the previous example a title is given and a description of the specific aims is also spelled out in the first part of the protocol.

Effect of Recombinant Human Chorionic Gonadotropin on Chemically Induced Rat Mammary Carcinogenesis

Specific Aims

Specific Aim 1. Evaluation of the efficacy of r-hCG in the prevention of mammary cancer.

The potential of hCG to inhibit the initiation of DMBA-induced rat mammary carcinomas will be evaluated in intact virgin rats following the protocol outlined in Fig. 2.

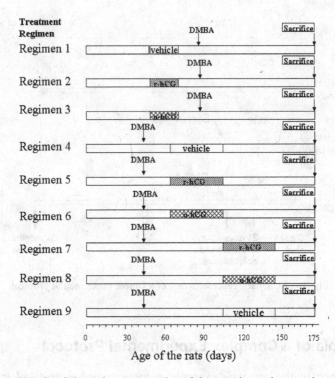

Fig. 2. Schematic representation of the experimental protocol.

Treatment Regimen 1: Sixty-five 45-day-old virgin Sprague Dawley rats will receive a daily intraperitoneal (ip) injection of 0.5 ml r-hCG vehicle for 21 days. Under Regimen 2 the same number of animals will receive 100 IU r-hCG for 21 days. Twenty-one days after completion of the r-hCG or vehicle treatments both groups of animals will be inoculated with a single intragastric (ig) dose of 8 mg DMBA/100 g body weight. Fifty animals from each protocol will be followed up by examination and biweekly palpation for detection of tumor development and determination of tumor growth rate and registered in an individual record's sheet (Fig. 3). Final tumorigenic response will be evaluated six months after administration of DMBA.

Five animals from each protocol will be bled at the time of DMBA administration, at the beginning of the administration of hCG or vehicle and at the end of the experiment. Whole blood will be obtained by cannulation

Group #	Animal #	hCG dose

Date →					
Rat Weight (mg) →					
Tumor #	Location				
1→					
2→					
3→					
4→					
5→					
6→					
7→					
8→					
9→					
10→					

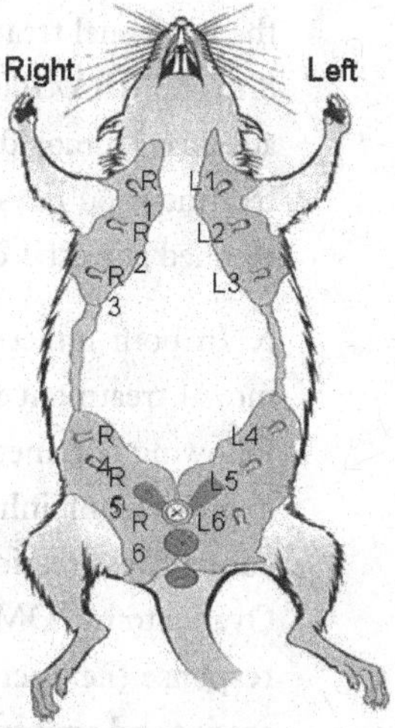

Fig. 3. Sample of an individual sheet for recording tumor evaluation. (Figure provided by Johana Vanegas, MD).

of the inferior vena cava and serum will be separated and stored frozen at –70°C for hormonal levels determination.

Specific Aim 2. Evaluation of the therapeutic efficacy of r-hCG on mammary cancer.

The tumoristatic and tumoricidal efficacy of r-hCG on early and advanced tumor development will be evaluated in two groups of intact virgin rats. For the effect on early tumor development the hormonal treatment will be initiated 20 days after administration of 8 mg DMBA/100 g body weight (Fig. 2, Regimen 3). The effect on advanced tumor development will be evaluated by starting the r-hCG treatment 60 days after DMBA administration, or when tumors had reached at least 1 cm in maximal diameter (Fig 2, Regimen 4).

a. The evaluation of mechanisms of action of r-hCG on early and advanced tumor development will be assessed in two groups of ovariectomized-hormonally supplemented virgin rats (Fig. 1, Regimens 6 and 7). These protocols will test whether this hormone acts directly on the tumor and normal mammary epithelial cells or has an indirect effect mediated by the ovary. Animals treated as indicated in Fig. 1, Regimens 1 and 5 will serve as controls. For the effect on early tumor development the hormonal treatment will be initiated 20 days after administration of 8 mg DMBA/100 g body weight (Fig. 2, Regimen 6). The effect on advanced tumor development will be evaluated by starting the r-hCG treatment 60 days after DMBA administration, or when tumors had reached at least 1 cm in maximal diameter (Fig. 2, Regimen 7).

b. In both intact and ovariectomized animals the effect of the hormonal treatment will be evaluated by determining in the tumors the following parameters: rate of tumor growth, tumor cell proliferation, expression of inhibin synthesis, estrogen and progesterone receptor content, expression of programmed cell death genes, and apoptosis. Ovariectomy, DMBA administration, and evaluation of tumorigenic response (i.e. incidence and multiplicity of mammary tumors, proliferative and apoptotic responses in tumoral and non-tumoral mammary glands) will be evaluated as detailed in the methods section.

c. Five animals from each protocol will be bled at the time of DMBA administration, at the beginning of the administration of hCG or vehicle and at the end of the experiment. Whole blood will be obtained by cannulation of the inferior vena cava and serum will be separated and stored frozen at -70^0C for hormonal levels determination.

Specific Aim 3. Evaluation of systemic effects of r-hCG. The effect of r-hCG treatment on the ovary and the pituitary will be evaluated by determining serum levels of hCG, estrogen, progesterone, inhibin, follicle stimulating hormone (FSH) and luteinizing hormone (LH) at six time points of r-hCG treatment in five animals from each one of the seven treatment regimens listed above.

Methods
Animals and carcinogen treatment

These studies will be carried out utilizing 350 virgin Sprague-Dawley rats, 200 intact and 150 ovariectomized (OVX), that will be maintained in an environmentally controlled clean air room with a 12-hour light/12-hour darkness cycle. They will be fed Purina Certified Rodent

Chow 5002 and tap water ad libitum. In Table 2 is listed the principal set of components needed for carrying on this experiment.

All intact animals and ovariectomized animals will receive a single intragastric (ig) instillation of 8 mg 7,12-dimethylbenz(a)anthracene (DMBA) (Eastman Kodak, Rochester, NY) per 100 g body weight when they reach the age of 82 days. Animals treated as indicated under Regimens 1 and 2 (Fig. 2) will receive a daily intraperitoneal (ip) injection of vehicle or 100 IU r-hCG, respectively. Injections will be applied for 21 days, starting when the animals reach the age of 45 days. Animals under Regimens 3 and 4 will receive a daily ip injection of 100 IU r-hCG starting 20 (Regimen 3), or 60 (Regimen 4) days after DMBA administration. Ovariectomized (OVX) animals will be treated with vehicle (Regimen 5), or 100 IU r-hCG for 40 days starting 20 (Regimen 6) or 60 (Regimen 7) days after DMBA administration.

All animals will be palpated biweekly for evaluation of tumor development. Palpable tumors will be sequentially numbered and measured for assessment of rate of growth. All animals will be sacrificed 24 weeks after the initiation of the experiment and all the mammary glands will be dissected for evaluation of gland development by morphological analysis in whole mount preparations and histological sections. Palpable and microscopic tumors will be dissected and processed for histopathological analysis, immunocytochemical detection of bromodeoxyuridine (BudR) incorporation, inhibin α and β subunits, apoptotic index, detection of estrogen and progesterone receptors by multipoint titration analysis of tumor cytosol fractions incubated with increasing concentrations of ^{3}H-estradiol 17 β, using the dextran–coated charcoal method, and DNA proliferation index by flow cytometry.

Ovariectomy and estrogen–progesterone pellet implant

Animals selected for Regimens 5–7 will be ovariectomized when they reach the age of 40 days old. This procedure will be performed in anesthetized animals through a midline incision in the skin of the dorsum. The dorsolateral muscles will be exposed by blunt dissection, and incised with sharp scissors; upon exposure of the ovaries, they will be tied with catgut and excised. The uterine cornua will be re-placed into the abdominal cavity; the muscle will be closed with 4-0 catgut fitted with an atraumatic needle and the skin will be closed with metal clips. The same anesthetized animals will have implanted with a needle-trocar in the subcutaneous tissue of the interscapular area an estrogen (1.0 mg)–progesterone (35.0 mg) pellet (Innovative Research of America, OH). The skin perforation closes spontaneously without the need of suturing. Pellets will be left in place for the length of the animal's life span. After surgery the animals will be placed on heated pads until they are fully recovered.

Assessment of mammary gland development in whole mount preparations

At sacrifice, the mammary glands of all the animals will be removed, attached to the skin pelt and fixed for 24 hours in 10% neutral buffered formalin. The mammary glands will be then dissected from the skin, defatted in acetone, stained with toluidine blue, dehydrated, cleared in xylene and mounted on glass slides with mounting medium. Mammary gland development and tumorigenesis will be evaluated under a stereomicroscope by counting the number of normal appearing terminal ductal structures, i.e. terminal end buds (TEBs), terminal ducts (TDs), alveolar buds (ABs) and lobules. The number of each specific type of structure present in thoracic and in abdominal mammary glands will be expressed as the percentage of the total number of structures counted in each gland, and the average of 50 animals per group will be obtained.

Evaluation of tumorigenic response

Whole mount preparations will be examined for microscopic lesions under the stereomicroscope. Every microscopic lesion identified will be photographed, dissected from the gland and embedded in paraffin for histological examination. Grossly palpable tumors will be fixed in formalin and processed for light microscopic examination. Final tumorigenic response will be evaluated 24 weeks post-carcinogen administration. All palpable and microscopic tumors will be fixed, embedded in paraffin, stained with hematoxylin–eosin and classified according to criteria published elsewhere. Tumor incidence will be assessed by the summation of all palpable tumors, microtumors identified in whole mount preparations and microscopic tumors identified in routine histological sections.

Evaluation of cell proliferation

Five animals from each one of the groups under study will receive an ip injection of bromodeoxyuridine (BudR) at a dose of 200 mg/kg body weight and two hours later the animals will be euthanized. The mammary glands will be fixed in 10% neutral buffered formalin (NBF), dehydrated through a graded series of ethanol, transferred to xylene and embedded in paraffin. Sections will be processed for light microscopy immunocyto-chemistry as described below. Based on our preliminary data, and assuming similar variances, we calculate that a sample size of five animals per group will provide 80% power to detect differences between means of approximately one standard deviation between DMBA, hCG+DMBA, DMBA+hCG, OVX-DMBA, and OVX-hCG+DMBA-treated animals, or between any three groups of interest. This number of animals maximizes the power to detect biologically meaningful changes in gene expression levels while minimizing animal costs.

Determination of tumor cell proliferation by flow cytometry

This procedure will be used for evaluating cell proliferation in all the tumors obtained from animals treated as under the regimens shown in Fig. 2. It is estimated that a total of approximately 200 tumors will need to be evaluated. The technique that will be used is a standard procedure in our clinical pathology laboratory.

Determination of tumor steroid hormone receptor content

The content of estrogen and progesterone receptors in DMBA-induced rat mammary tumors will be evaluated by multipoint titration analysis of tumor cytosol fractions incubated with increasing concentrations of ^{3}H-estradiol 17 β, using the dextran-coated charcoal method. These procedures will be performed according to the standards established for the assessment of estrogen receptors in human breast cancer established by the E.O.R.T.C. Breast Cancer Cooperative Group at the Antoni van Leeuwenhoek-Huis workshop in Amsterdam. The evaluation the content of estrogen and progesterone receptors will be performed in all the tumors obtained from animals treated with DMBA, as shown in Fig. 2. It is estimated that a total of approximately 200 tumors will need to be evaluated. The technique that will be used is a standard procedure in our clinical pathology laboratory.

Specific details of the material needed, hormone preparation and dosage are listed in Tables 1 and 2.

Table 1. Material needed for the experiment.

Material or drug	Description of usage
DMBA-Eastman-Kodak: 5 grams.	Corn Oil: 250 ml At a dose of 8 mg/100 gram body weight. DMBA will be suspended in corn oil at a concentration of 10 mg per 0.5 ml corn oil. If the animals weigh an average of 125 grams they will receive 10 mg (0.5 ml suspension). Injection of a total of 455 animals will require 4,550 mg (5 grams DMBA) dissolved in 250 ml corn oil.
Recombinant hCG: 136.5 vials of 250 micrograms each.	Every vial contains 250 micrograms = 5000 IU. At a dose of 100 IU/day for 21 days, administered to a total of 325 animals (100 IU × 21 days × 325 animals = 682,500 IU [34125 micrograms).
Gastric cannulas	(for injecting DMBA)
Syringes	10 ml: 140 (for bleeding) 3 or 5 ml: 40 (DMBA administration) 1 ml: 956 (for injecting hCG or vehicle if every syringe is used once for injecting 10 animals only, without refilling)
Blood separator vacutainers	140
Ear punch marker	Animal Sheets
Anesthesia	Dry ice (for DMBA administration) Ketamine (90 mg/kg bw) + Xylazine (10 mg/kg bw), ip for bleeding and tissue dissection
Gloves	
Hair covers	
Masks	

(Continued)

Table 1. (*Continued*)

Material or drug	Description of usage
Disposable gowns	
Shoe covers	
Caliper	(for measuring tumors)
Balance	(for determination of animal weights)
Surgical material	3 sets — Each set should consist of:
	1 large pair of large scissors
	1 pair of iris scissors
	1 pair of forceps (medium size)
	1 pair of fine forceps
	1 scalpel handle
	Scalpel blades
Wooden boards	
Pins	
Sharps containers	
Bags for animal disposal	

Table 2. Treatment regimens using recombinant hCG.

Group	Treatment protocol	Schedule/Dose	hCG Treatment	
			Start	End
Regimen 1	65 animals injected with placebo daily for 21 days	65×0.1 ml $\times 21 = 27.3$ vials	3/8/99	3/28/99
Regimen 2	65 animals injected with 100 IU/ r-hCG daily for 21 days	$65 \times 100 \times 21 = 136,500$ IU $= 27.3$ vials	3/8/99	3/28/99
Regimen 3	65 animals injected with 100 IU/ r-hCG daily for 40 days	$100 \times 65 = 6500 \times 40 = 260,000 = 52$ vials	5/10/99	5/30/99
Regimen 4	65 animals injected with 100 IU/ r-hCG daily for 40 days	$100 \times 65 = 6500 \times 40 = 260,000 = 52$ vials	5/31/99	6/21/99
Regimen 5	65 animals injected with placebo daily for 40 days	$100 \times 65 = 6500 \times 40 = 260,000 = 52$ vials	5/10/99	5/30/99
Regimen 6	65 animals injected with 100 IU/ r-hCG daily for 40 days	$100 \times 65 = 6500 \times 40 = 260,000 = 52$ vials	5/10/99	5/30/99
Regimen 7	65 animals injected with 100 IU/ r-hCG daily for 40 days	$100 \times 65 = 6500 \times 40 = 260,000 = 52$ vials	5/31/99	6/21/99
OVX-Regimens 5,6,7			3/3/99	
DMBA injection			3/17/99?	
End Experiment			4/19/99	9/20/99

Total r-hCG needed: 682,500 IU = 136.5 vials.
1 vial = 250 mcg = 5,000 IU.

172

Chapter 8

The Ten Fatal Mistakes

1. Underestimating the Mentor

The relationship between the Mentor and the scientific apprentice consists of several phases that move between mutually high expectation, admiration, doubt, acceptance, condescension and finally, recognition. These phases occur naturally and once we are made aware of them, we find they can be part of any human relationship.

One of the golden rules the scientific apprentice must follow is that they must at all times be cognizant of his or her mentor's experience, and maintain a sense of respect for their opinions. It is also important to identify the reasons for the mentor's criteria and how the mentor feels about a variety of subjects. This is not to say that the mentor is always right, but the scientific apprentice must learn to discuss any contentious issues and to address disagreements with respect and courtesy. It is important that these conversations be initiated in the spirit of scientific discussion and not out of emotional reaction. The interesting part of the mentor–mentee relationship is that the mentor may also experience upheaval on his or her end of the exchange. In this case it is the mentor's experience that must guide a steady flow of energy and strength between him or herself and the scientific apprentice.

Within the phase of high expectation is the process of finding the right mentor. Scientific apprentices tend to look to mentors as a sort of key that will open locked doors, ultimately securing their success. Even when comparing their ideal mentor with all the other possible choices, the fact of having been selected or chosen by a mentor sweetens the scientific apprentice's opinion of that researcher. Because of this initial bias, I strongly suggest that the scientific apprentice search as much as possible, discuss their thoughts with other scientific apprentices and older scientists, compare, and ask as many questions as possible on the mentor's fairness, tolerance for deadlines, and stringency of criteria for accepting data. The scientific apprentice should also be extremely familiar with the mentor's work and his or her motivations, as well as have a sense of who they are in their private life. It is important that once armed with this basic knowledge the scientific apprentice has more that one conversation with the mentor to discuss the pros and cons of the approaches to they wish to follow, and respectfully discuss those issues that might raise a nagging concern and could unconsciously create a reservation about the mentor.

Every human relationship sees its fair share of upheaval and doubt, and the mentor–mentee relationship is no exception. The chemistry between the mentor and scientific apprentice must be right from the start. If there is any doubt in the scientific apprentice's mind about a mentor's fairness or experience working in an advisory capacity with young scientists, then the apprentice should seek elsewhere. Nor should an apprentice simply choose the first person available. Likewise, a good mentor will decline to mentor someone who they have reservations about.

It is vital that a mentor have experience mentoring and this should be an important consideration when deciding whom to approach with the question of mentorship. But how is one to recognize the best choice? When embarking on this critical, career-shaping search, a scientific apprentice is well served by knowing what to look for, and understanding a little bit about the mentor's experience as well.

Over the course of my career I have mentored over 50 scientific apprentices, which has given me a good sense of who will be a good fit in my lab and who will not. Several years ago a young man approached me, interested in working in my lab, with the hopes of going onto postdoctoral training. He was a very bright and highly motivated individual who had recently graduated from a first-rate university in the Midwest. He had excellent grades and even a couple of publications under his belt. After reviewing his CV, recommendation letters, and hearing his presentation during an official interview, I offered him the position. He did not accept my offer, however. Another laboratory offered him a similar post, the difference being that the other lab had suggested that he would be able to move ahead at a faster rate by publishing more papers and heading more projects than I was willing to promise. Three months later he came back, a little sheepishly, and asked me if the position I had offered him was still open. Lucky for him, it was, and this time he didn't waste a minute before taking me up on it. In the months that followed, the fact that my lab had in his mind, remained a suboptimal choice became evident. He seemed restless and disengaged. Even though I had done my best to give him what he needed, he wasn't happy working with me. Even though the situation had worked out well for this young man, what went wrong? In accepting him, I made the mistake of failing to notice that this bright individual had accepted a situation that was not optimal for him. After a year with me he left, and went on to a different lab that better suited his interests. I'm happy to say that he has done well since then and continues to follow his passion for science.

The lesson I learned was that a mentor, as much as the mentee, must pay attention to what the apprentice's aspirations are, as well as what they expect to gain from the relationship. A failure to do so results in mutual frustration, even negative feelings. In this case, no long-term damage was suffered on either of our parts. The experience did teach me, however, not to be accepted as a second, or "fall back" choice, not only for my sake, but for the apprentice's. It's a policy I have followed ever since.

With this young man I was unable to foresee natural phases of expectation, admiration and doubt that are a natural part of the process of beginning an apprenticeship. At the time of hiring he was in the doubt and condescension phase, what I now know is a clue that the match was not right for either of us.

Over the long course of interaction between the mentor and the scientific apprentice the doubt phase is the one in which the scientific apprentice suffers most. For those preparing their doctoral thesis this generally occurs at the end of their second or third year, but postdoctoral fellows are not exempt. This is the period when doubts begin to surge in the scientific apprentice's mind about their mentor's ability to put them on the right experiments or help him or her achieve the expected number of publications. It is in this period when the mentor must play an important role by reaffirming the apprentice's confidence and help them reevaluate their work, while at the same time being extremely critical and firm, always underscoring that a career in science is not simply that, but a true vocation.

All those who have begun their doctoral work in my lab have finished successfully. However, a few of my postdoctoral fellows through the years have decided not to pursue scientific research as a career. Some are working successfully at pharmaceutical companies, others are practicing medicine, and one went so far as to change careers entirely. This goes on to show that the mentor's role is not to dictate someone's interests or direction, rather to help cultivate the desire that exists within each individual.

Part of the mentor's function is to help orient the scientific apprentice towards the path that is best suited for them. Bright, over-ambitious people are often not the ones best suited for work in scientific research. It is those who find joy in studying, acquiring knowledge and making discoveries who will truly find their home in research, not those whose desires in life center on success for success' sake, or with expectations of high dividends for their efforts. This is an important distinction; the phase of doubt and condescension toward the mentor are a dangerous moment in the relationship between the apprentice and the mentor. The scientific apprentice's confidence has grown because his or her observations are not only confirming the mentor's hypothesis but are also contributing a new vision that could potentially be a new line of research that might give this new budding scientist an edge. It is from this vibrant sense of nascent ability that a feeling of competition with the mentor can easily arise. The mentor must be aware of this and learn to manage and give breathing space to those scientific apprentices that want to fly solo into the world of science. But the approach must be delicate, compassionate and full of understanding for this new situation. The mentor must in this moment maintain a clear vision of his or her mission while remaining generous and patient. During this period the mentor must talk not only of the scientific findings, but also openly discuss with the apprentice his or her immediate future, help them measure the strength of the work and advise them on how to proceed towards independence. When handled well the scientific apprentice not only grows, but a lasting relationship is established.

2. Sloppy Work

Those who seem to have an innate ability to make plants grow, flowers bloom and seeds sprout are said to have a "green thumb." In the research laboratory it is expected that experiments be well planned, and that results be meaningful and conducive to original observations, all in the service of publications and the development of hypotheses. This is what it is expected. Unfortunately not all the experiments are well planned and sometimes even the best ones do not confirm a particular hypothesis or data are not conducive to success. The scientific apprentice must face the fact that planning and discussing an experiment from its inception are crucial parts to scientific research and that there are no short cuts. Experiments do not simply flourish — they have to be cultivated. There is a difference between a well-planned experiment that does not demonstrate the hypothesis, and suboptimal or sloppy work, what in laboratory jargon we call a "dirty experiment."

What constitutes sloppy work? Sloppy work goes beyond neatness and physical order. When an experiment has not been planned well, meaning it has not been designed with the adequate statistical relevance to see an effect, adequate controls, either positive or negative, have not been included, the number of parameters to be studied were not properly estimated (meaning too many parameters or too few), all result in waste of the material collected and supplies used. But there also additional mistakes that make experiments sloppy and ultimately fail, such as methodology that has not been properly tested before hand, use of expired reagents, or inadequate timing (such as beginning a daily treatment before a holiday), faulty sample collection, or an apprentice working alone in the lab when they in fact might need assistance, all skew or nullify the results. Therefore, the scientific apprentice must meticulously plan and execute all experiments. One cannot approach experimentation with the idea that if something goes wrong it can simply be repeated. Well-executed experiments with the best of results need to be repeated for confirmation or validation. But that conducting a second or third validation is different than simply repeating because the first attempt failed due to an error of oversight. When this is the case the second attempt, if done well, only counts as the first.

Sloppiness can carry through to every single aspect of research, such as incomplete records in the protocol book, inadequate preparation of solutions or poor calibration of instruments. The details are many and something that the scientific apprentice must focus on from very early on.

What are the consequences of sloppy lab work? The consequences can range from outright dismissal to a protracted apprenticeship with the subsequent lost of confidence from the mentor. The seriousness depends on the magnitude of the mistake; if radioactive material spills because of careless handling, this is doubtlessly a serious incident that can jeopardize the whole laboratory and even lead to closure of the lab. Carelessness at the level of a single contaminated culture plate is of a different magnitude than contamination of an entire batch of culture media in that more than one member of the lab, and most likely more than one experiment, are affected. While the consequences of certain errors vary, all thoughtless mistakes have an effect on everybody involved.

If a scientific apprentice is unable to correct their deficiency in paying attention to detail and protocol early on, or never achieves a level of comfort when it comes to bench work, they may be forced to realize that the research laboratory, or at least the "wet" part of it, is probably not the best environment for them. If a mentor does not take a proactive action to correct deficiencies that most of the time are easily corrected, failed experiments and botched results are almost inevitable.

3. Falsifying Data

Falsifying or altering data is a serious act of misconduct in science that results from a misunderstanding of what scientific research is all about. The scientific apprentice must strongly believe in the *process* of doing biomedical research, that we do this work because we honestly seek the truth and understanding about the mechanisms of life. Therefore to manipulate any data is the negation of the principle that defines us as scientists.

Although there are never valid reasons for falsifying or altering data, there are multiple occasions that may make it tempting. Among them are lack of results, deadline pressure, competition or to satisfy the mentor's hypothesis.

It is the mentor's responsibility to ensure that none of these situations take place. The mentor must maintain continuous contact with the apprentice and all staff, and evaluate on a weekly basis the progress of the experiments and discuss data as it unfolds. If a pattern of frequent communication and contact is embraced it is unlikely that any scientific apprentice will dare to alter data, nor will they feel the need to. Open communication fosters a supportive environment that minimizes a lot of the anxiety that leads people to falsify. However, if the apprentice is left to fend for themselves in a laboratory where the number of students and staff exceeds the supervisory capacity of a single mentor, it becomes the kind of environment where a developing scientist might feel compelled to show something meaningful, even if it isn't true. Group pressure and the competitive environment of the laboratory can also be contributing factors. It is therefore the responsibility of the mentor to recognize how many apprentices he or she can properly mentor. The responsibility also falls upon the shoulders of granting agencies that provide the fellowship funds to scientific apprentices in that they are tasked with evaluating whether the mentor has the experience and resources to properly assist a scientific apprentice with a given project. They do not want to support a proposal that is likely to fail because the mentor is spread too thin, or has insufficient experience. Unfortunately this is a serious mistake that some grant reviewers make when they closely scrutinize the scientific apprentice's fellowship application but do not apply the same rules to the mentor.

Special attention should be paid to those scientific apprentices who, for their own reasons, feel the need to get the approval of the mentor by forcing data in the direction of the preferred hypothesis. Again, it is the mentor that needs to be careful to not show preferences in the laboratory; all scientific apprentices and staff must receive the same level of respect and treatment from the mentor. He or she must be extremely careful in not overvaluing the work of one scientific apprentice over another. In most cases it is in the interaction between the mentor and the scientific apprentice (*individual exchange* interaction as discussed in Chapter 6) that this kind of favoritism is easily spotted.

When the mentor has even the minimal suspicion that the data collected are forged, adequate measures like blind experiments and randomized samples must be run to positively confirm or deny the suspicions of forgery. This, however, needs to be done without destroying the working relationship, yet put significant pressure on the apprentice to correct their attitude. In addition, the mentor must directly or indirectly discuss this transgression with the scientific apprentice. A problem of this kind needs to be resolved before the time comes for the scientific apprentice to move on from the laboratory.

4. Jealousy

This section is not intended to be a treaty in psychology but to illustrate how jealousy plays a role in the relationship between the scientific apprentice and others, and more importantly, with the mentor. The basic dictionary definition of jealousy is "an emotion with negative thoughts and feelings of insecurity, fear, and anxiety over an anticipated loss of something that the person values, such as a relationship, friendship, or love." In life, jealousy often consists of a "combination of emotions such as anger, sadness and disgust." The reasons jealousy plays an important role in the mentor–mentee relationship are often based on expectations that the scientific apprentice has on what he or she will accomplish during training. Those with unrealistically high expectations are the ones that will suffer the most and be jealous of anybody else that might take the spotlight away from them, even for a moment. They need the mentor's continuous support and if they perceive, whether valid or not, that their position in the laboratory is threatened, a sensation of uselessness consumes them. This internalization of exterior factors beyond their control can make them very depressed and decreases their productivity. Likewise, the same person may feel exuberant and display excessive amounts of energy if they perceive that they are regaining the center stage. A scientific apprentice that wants to be the only person recognized might show this subtly, or be a nightmare for the mentor.

While these descriptions might seem extreme, these personalities are more common than one might expect. Fair treatment of everyone in the lab will probably mitigate jealousy; however, the mentor must detect these subtleties and refrain from forming preferences, especially in the presence of such people. In my experience these types of scientific apprentices are difficult to deal with, but if the mentor is fair in the treatment of all the members of the lab and the scientific apprentice only expresses jealousy as a normal and not a pathological process, the situation can be managed.

5. Vandalism

Fortunately there are not many cases, at least known or reported, in which a scientific apprentice has massively vandalized and destroyed a lab, behavior that in most cases is an extreme manifestation of rage. There are however, small acts of vandalism that can result from the scientific apprentice's frustration toward the mentor or the group that may go undetected at the beginning, but could eventually be serious enough to call for drastic action against the perpetrator. The mentor must be aware that frustration, doubt or even jealousy, can trigger a willingness to damage the mentor by perpetrating small, perhaps unconscious, acts of vandalism such as inadequate care of equipment, thoughtlessly ordering unneeded material, disposing of perfectly good reagents or carelessness in the use of daily tools. On some occasions the period of doubt and condescension towards to mentor produces no material damage, rather, verbal expressions of discontent and negative remarks toward the mentor or his or her institution which is also, in its own way, a form of vandalism. It is important that the mentor be familiar with of each scientific apprentice's personality and knows how each of them responds to stress. The scientific apprentice should also keep in mind that he or she is not exempt from feelings of anxiety, frustration and even violence, and that the laboratory is a place made up of many types of people and therefore a potential setting for these feelings to emerge. It is the responsibility of the mentor to maintain a safe and healthy atmosphere where only the best of each scientific apprentice can rise to the surface.

6. Cynicism

A mentor is expected to operate in accord with the overall philosophy that defines their work, and as a consequence it is expected that the scientific apprentice, to a certain degree, adopt the same approach. I strongly believe that the mentor must embrace science as a vital part of his or her life and that the pursuit of the truth should fuel the idea behind his or her research efforts. But how many mentors practice this philosophy? I cannot clearly indicate a number nor can I accurately count how many of the scientific apprentices I've worked with have had this motivation in mind. This is especially true in a multicultural society where values and ideals are derived from multiple, often opposing, perspectives, it can never be perfectly clear whether the principles that guide a mentor's life will be the same ones that guide the scientific apprentice's life.

Despite differences in personal belief and opinion, there is one singularly important pitfall that in my experience as a mentor I have become apt at detecting early on: cynicism. Because of its very nature cynicism is not easy to detect — often it is obscured or confused with objectivity. Cynicism, however, is not objectivity, rather it is a disregard of the basic principles that good science and honest work are the driving force and leitmotif of a scientist's life.

7. Postponing Deadlines

Deadlines are more than just points on a calendar — they are the tempo that marks our productivity and the accomplishment of our goals. Deadlines are not arbitrary in that the mentor does not close his or her eyes, spin around, and let their finger land on whatever date or time it will. They are determined by meeting schedules, grant reports and renewals, and new applications. Most deadlines are established with the level of a project's maturity or complexity in mind. To be a scientist in this era requires a very organized mind and the ability to plan five to ten years in advance. It is pure fallacy to believe that one can generate a manuscript in a single day, or week, or that a project rises from a bout of inspiration. The scientific apprentice must accept that the pace of his or her work is determined by deadlines and must learn to respect and live with them.

A scientific apprentice that does not respond to deadlines can create not only a serious problem for his or her own career and advancement, but also be a disruptive element to the team. The mentor must be aware that deadlines are easier for some people than others and it is his or her function to work with these individuals in order to re-align these differences in perceiving time. The scientific apprentice who struggles with deadlines needs to work toward organizing their priorities and incorporate the importance of deadlines into the daily practice of their work.

8. Underestimating Your Peers

In a multicultural society we are often faced with points of view very different from our own. It is the mentor's obligation to understand the various cultural patterns in his or her lab, and be sensitive to the cultural background of any incoming scientific apprentice, regardless of where they come from. If, for example, a mentor is interested in taking on a scientific apprentice from China, the very least the mentor can do is to learn about the candidate's culture and take the opportunity to learn more during the interview before even offering them a position in the laboratory. The same applies to any other scientific apprentice coming from those countries often referred to as "Hispano-American." While many countries such as Cuba, Venezuela, Ecuador, are all Spanish-speaking countries, they are by no means culturally identical. To believe that a scientific apprentice coming from Mexico has the same cultural heritage and way of life with one coming from Chile can be a source of misunderstanding and frustration. In many ways we have grown used to the idea of a multicultural society and today are fortunate enough to be exposed to various customs and traditions. Despite this openness, even the best and most open-minded mentor could face problems in the laboratory when people of different cultural backgrounds are put in the same work environment. One of the most frequent problems in my experience is the underestimation of one group by another. This friction in the workplace is subtle most of the time yet needs to be identified because it has the potential to jeopardize the entire team's morale. The same applies to language usage. If there is a preponderance of one group over the other in a lab, the mentor must reinforce the use of English at all costs to avoid derogatory attitudes amidst the group. These kinds of biases are resolved by energetic and decisive action on the part of the mentor and by sustaining a zero tolerance policy against gender, cultural, religious and other forms of discrimination.

9. Overconfidence

Many years ago when "the skies were friendly" and the events of September 11 were beyond what we thought was possible, I asked one of the airline attendants on my flight whether I could take a look inside the cockpit. I had always been curious about what went on in that tiny cabin that allowed 747s to go soaring through the air. Even more than the complex panel of precision instruments and gauges, I was most impressed by one single thing: both the pilot and co-pilot, seasoned professionals, were constantly checking and re-checking the operations manual. Despite a combined 34 years of experience and countless flight hours, they continued to verify the steps of the flying process, and to do so on every flight. Overconfidence breeds error and in too many cases has been the source of grave disasters — in the air as well as in the lab. Following a checklist, no matter how many times you've done something, is probably the best way to ensure a good performance, not only in flying an airplane but in most of the things that we do in science.

Confidence is an outstanding quality, overconfidence, however, brings that person closer to arrogance and foolishness which is always bad for scientific research. Why is overconfidence a fatal mistake? In my personal experience, overconfidence decreases mental awareness, or alertness, that as scientists we must maintain in order to ask the right questions of our findings and interpretations. When we become overconfident we are prone to making the wrong judgment about our own ideas and even placing our interpretations above all others. Overconfidence compromises our ability to listen.

When the scientific apprentice is preparing a lecture, for example, the lecture's primary mission is to transmit a message that the audience will understand. As Pasteur indicated many years ago, first you need to convince yourself that your reasoning is correct, then you need to convince others, and third you need to convince your adversaries. Therefore, if you are overconfident you will loose the edge that being aware of your hypothesis' pitfalls can bring.

10. Indifference

I wanted to leave the problem of indifference for last because this is probably the most dangerous of all the pitfalls that I have discussed above.

Indifference means that the basic itch that drove the scientific apprentice to science is lost or mitigated to a point that the challenges are nonexistent, or so powerful that the scientific apprentice feels that they are not worth overcoming.

The causes of indifference are so varied it is difficult to pinpoint when it starts. A mentor who drives his lab hard and with a lot of enthusiasm and energy could be a positive source of inspiration for one scientific apprentice, but to another that same mentor could be too much to bear. This could happen without the mentor realizing that he or she has quenched the scientific apprentice's fire by overwhelming them. This dynamic is not so different than what happens in households headed by parents who demand high levels of achievement. Unless the parents know how to balance high expectations with encouragement and support, they risk pushing their kids to the opposite extreme. In this regard the mentor and the parent are not so different. If pushed too relentlessly the scientific apprentice may become indifferent to what is meant to be incentive. If indifference festers to the point of cynicism, the scientific apprentice has reached a perilous state.

It has been my experience that the mentor must get to know their scientific apprentices and help them understand their limits. To succeed in science, apprentices must learn that it is not geniality that counts but perseverance and a disciplined mind. Surely it helps if the scientific apprentice is a genius, but not as much as discipline and sheer will. Therefore, the mentor must insist on these two important components of scientific life by helping the scientific apprentice remain perseverant and focused on the problem at hand, as well as intellectually disciplined and attuned to the rigors of practical bench work.

There are other factors that can make the scientific apprentice indifferent. The most prevalent is the lack of job opportunities and funding for research. In biomedical sciences, both aspects are linked and when the scientific apprentice sees that the prospect of getting a grant is at an all-time low, the sense of having no future career can lead the scientific apprentice to start loosing motivation and interest, which may end in indifference. The scientific apprentice may then start searching for a new way of life that is not as difficult to achieve as a scientific career. The mentor plays an important role in avoiding discouragement, but there is a limit to what he or she can do if the prevailing economic situation puts even the mentor's job in peril. Mentor's should discuss the problem of funding and job placement openly with scientific apprentices and instill in them a sense of self-reliance and a willingness to search for possibilities in avenues that may initially appear to have no value.

In the end, we as a society are responsible for maintaining the next generation of scientists, as well as writers and artists, for the wealth of humanity. These and other enterprises, such as those mentioned, are probably not destined to provide great economic dividends in the short-term, but are the core of our human social network, which is why society, through government organizations and the like, have been tasked with making sure this network continues to grow.

Suggested Readings

Abelson H. Keeping the net stupid. *American Scientist* **96**:503, November–December 2008.

Cassidy DC. Born unto trouble. *American Scientist* **93**:372–374, July–August 2005.

Disputed definitions. *Nature* **455**:1023–1028, October 23, 2003.

Fitch V. The bomb's hometown. *American Scientist* **93**:369–370, July–August 2005.

Gliboff S. Love, death and Darwinism. *American Scientist* **97**:68–69, January–February 2009.

Harpham G. Science and the theft of humanity. *American Scientist*, Vol. 94, July–August 2006.

Hopkin K. A mind apart. *The Scientist*, p. 56, June 2008.

Hunter P. Educating Oxbridge. *The Scientist*, p. 50, December 1, 2003.

Inspection Science. *Science* **299**:1281, February 28, 2003.

Kelves BH. Edison —— The view from Menlo park. *Science* **282**:1997–1998, December 11, 1998.

Latour B. From the world of science to the world of research. *Science* **280**:206–209, April 10, 1998.

Marshall E. A big deal but a complex hand to play. *Science* **280**:196, April 10, 1998.

More questions about research misconduct. *Science* **297**:13, July 5, 2002.

Nisbert MC and Scheufele DA. The future of public engagement. *The Scientist*, p. 38, October 2007.

Park P. Celebrity ethics. *The Scientist*, p. 53, December 1, 2003.

Ploegh HL. Viral strategies of immune evasion. *Science* **280**:248–253, April 10, 1998.

Power S. Honesty is the best policy. *Time Magazine*, October 29, 2007.

Redman BK and Merz JF. Scientific misconduct: Do the punishments fit the crime? *Science* **321**:775, August 8, 2008.

Rouse J. Counter-revolutionary Kuhn? *Science* **288**:1755–1756, June 9, 2000.

Rutkowski JL and Feverstein GZ. Academic medicine to the rescue. *The Scientist*, p. 28, June 2008.

Secko D. Rare history, common disease. *The Scientist*, p. 38, July 2008.

Small MF. A women's curse? *The Sciences*, pp. 1–52, Jan/Feb 1999.

Stent GS. A picaresque genius. *Nature* **403**:827–828, February 24, 2000.

Stewart CN, Jr. and Edwards JL. How to teach research ethics. *The Scientist*, p. 27, February 2008.

Swan A. Open access and the progress of science. *American Scientist*, Vol. 95, May–June 2007.

Titus SL, Wells JA and Rhodes LJ. Repairing research integrity. *Nature* **453**:980–982, June 19, 2008.

Two former grad students sue over alleged misuse of ideas. *Science* **284**:562–563, April 23, 1999.

Vinge V. Win a Nobel Prize. *Nature* **407**:679, October 12, 2000.

Warren C. If at first you don't succeed... *American Way*, December 1, 2002.

Wiley S. My favorite fraud. *The Scientist*, p. 29, September 2008.

Wilson A. The tenure track quest. *The Scientist*, p. 59, December 1, 2003.

Wolfe AJ. Innovators and iconoclasts. *American Scientist* **97**:77–78, January–February, 2009.

Chapter 9

What is Expected of the Scientific Apprentice?

Although *every* scientist is an apprentice of science, because learning never ends, there is no doubt that a specific period of formation exists which I refer to as the apprenticeship of science. This chapter is directed to those beginning their tenure in scientific apprenticeship.

Good Work, Technically Perfect and Well Planned

It is in this formative period that certain standards need to become second nature, among them, how to produce good laboratory work that is technically perfect and based on well-planned experiments. These are the basic skills, or habits, that the scientific apprentice must acquire. Once they have been incorporated into daily practice, they become a permanent part of the scientific apprentice's muscular memory. Technical acuity and precision give a scientist tremendous confidence when facing the complex problems inherent to laboratory research. In these days of rapidly changing technologies, the more the scientific apprentice learns at the bench, the better they will be able to plan and execute good experiments and understand the full spectrum of scientific endeavor when their turn comes to be a mentor later on in life.

I have already discussed the importance of careful experiment planning and the protocol book, which are essential to good practice. It is expected of every scientific apprentice that data be properly collected and clean, in other words, trustworthy, enough to follow through to the next steps — interpretation and elaboration.

Critical Analysis of the Data Collected

Following data collection is analysis and interpretation of the results. This process requires several steps that involve going back and forth with the mentor.

In the first step the scientific apprentice describes the data and verifies that they have been properly collected. If there are variations or discrepancies in the study groups, or if the results are for some other reasons not acceptable, this is the time to go back and perform the experiments again to make sure that all the experimental conditions are properly considered. It is crucial that the mentor analyze the data alone and in the presence of the scientific apprentice, and discuss all the possible variations that can be made to the experiment, as well as discuss the pros and cons of each approach. Even the most experienced mentor should take the time to analyze the data and prepare, in detail, how to proceed, because at this point in the process the mentor cannot afford to be overconfident.

In the second step of data analysis the mentor will ask the scientific apprentice for their interpretation of the data. This step has an added level of complexity because the scientific apprentice's intellectual reasoning is on the table. The mentor should be circumspect in his or her own opinion regarding the interpretation of the data; there will be enough time to add later on if needed. The important thing at this step is for the mentor to guide the scientific apprentice in the right direction by suggesting additional reading, orienting a literature search, or by providing a few glances in the direction that the data are going, but that the scientific apprentice may not yet see clearly. It is possible that the scientific apprentice is already thinking ahead, perhaps sees the connections within the results sooner than the mentor had anticipated, in which case the mentor can go to the next step. This part of data interpretation, however, is when the mentor must show his or her intellectual guidance by suggesting possible avenues of interpretation, but in a more general rather than specific way because this is the opportunity for the scientific apprentice to read and compare their data with what is already known. It is also important that the mentor not provide, at this stage, any of his or her previous publications that might help with the data interpretation, because it is likely that the scientific apprentice will base his or her first interpretation on the published work provided, and miss the opportunity of searching the literature and coming to an original conclusion.

The third step requires that the scientific apprentice present a more refined set of data with an elaborated interpretation. It is in this step that the mentor must challenge the interpretation and ask for alternatives. If the scientific apprentice has done their homework, chances are they will be able to think on their feet. If at this point the scientific apprentice needs a push in the right direction, this is the time to provide some insight. It could also happen that the scientific apprentice comes with a genuine, novel, or even better idea than the one conceived by the mentor. Here the biggest test for the mentor is to listen without omitting any opinion from consideration and take some time to mull over the ideas posed to him or her. If there is something flawed, a new meeting will be set up to discuss alternatives, weaknesses, strengths, or even better, to congratulate the scientific apprentice for this new path.

The fourth step provides yet another level of communication between mentor and scientific apprentice. The data has been collected, analyzed and an interpretation leading to a hypothesis is emerging. At this stage the scientific apprentice is eager to determine whether the conclusions and hypothesis generated are valid and it is to this end that further experimentation should be carefully discussed. This is the scientific apprentice's opportunity to demonstrate how what they have learned up to that point will be able to help them generate the experimental protocols to demonstrate their hypothesis. It is at this moment that the mentor should provide the scientific apprentice more breathing space by asking them to present their findings and analysis to the group. This will likely be the first time the scientific apprentice's ideas will be submitted to the scrutiny of his or her peers.

The fifth step marks a critical period in the scientific apprentice's life. The exploratory phase is now over and the scientific apprentice now has their own hypothesis to prove. The mentor will have given permission to concentrate on a set of experiments that will help determine whether the hypothesis is correct or not. Whether the scientific apprentice has what it takes to be a scientist is often revealed at this stage. Every skill and bit of effort must be put into action to carry out the idea to its end. This step may last months, or a couple of years, but the end result will be publications, a doctoral thesis and a path or paths of knowledge that did not previously exist, but do now because of the scientific apprentice's work. The mentor must be there all the time and if both are congenial and mature people, a lasting relationship of affection and respect will link them forever.

Drive for Excellence

The need to strive towards excellence in the research endeavor is a message that needs to be constantly reinforced in the scientific apprentice's mind. While the period of scientific apprenticeship is only the beginning of a long career, it is the moment to establish standards of how to proceed in the near future. Excellence in science is defined by one important quality: a sound hypothesis that reflects an original thinking process. *There is no magic formula for excellence; it is a combination of the scientific apprentice's innate intelligence, experience gained from working with the mentor, an ability to grasp old as well as newly emerging concepts, and the ability to connect ideas that nobody has connected before.* This latter quality is probably the most difficult because more than anything else it requires a certain kind of intuition, not to mention strength and will, to formulate new avenues of research that nobody has forged before. This instinctual foresight is what, in the end, sets one scientist apart from another. Most scientific apprentices are able to see one or two connecting points and reach a level of comfort because they have proved they can do something significant. Their contribution might be well received and may even make them part of the emerging scientific community. Few, however, see more than the simple connections and are able to formulate biological laws and principles behind their findings which can forever alter the biomedical sciences.

So, is any one scientific apprentice better than another? In my view, no, but those who are able to think beyond the obvious, see the overall picture and foresee the implications of their findings will in the end play a bigger part in advancing science. This kind of out-of-the-box of thinking, however, opens the way to an arduous path. A principle that may have been conceived clearly, fluidly even, can take years, or even decades, to be proven true.

Clear Verbal Communication

From early on the scientific apprentice is called to the podium to discuss papers, or present preliminary or final data to be discussed and exposed to the scientific apprentice's peers. Unless a scientific apprentice has experience on a debate team or something of the like, this kind of public speaking requires preparation. What's the difference between giving a lecture and talking about the contents of a lecture? Audiences have notoriously fragile attention spans, therefore it is important that the speaker reach the listener and that the listener understand and retain the message being delivered. This requires a significant amount of preparation, and while a good speaker may sound off-the-cuff and spontaneous, chances are they have prepared well. The scientific apprentice must consider that each minute at the podium requires at least one hour of thorough preparation. Many speakers confuse entertainment with public speaking; scientists do not need to entertain an audience, but they should captivate them with a world that was previously unknown to them.

Suggested Readings

Cohen A. When the field is level. *Time Magazine*, July 5, 1999.

Communicating effectively in departmental meetings. *ASCB Newsletter*, pp. 19–20, April 2000.

Edwards TM. Harvard vs. the school of hard knocks. *Time Magazine*, June 21, 1999.

Glanz J. Human rights fades as a cause for scientists. *Science* 282:216–219, October 9, 1998.

Harding S. Women, science and society. *Science* 281:1599–1600, September 11, 1998.

Helmuth L. NIH under pressure boosts minority health research. *Science* 288:596–597, April 28, 2000.

Kreeger KY. Scientist as teacher. *The Scientist* 14(18):30, September 18, 2000.

Lawler A. Wanted: The ideal science advisor. *Science* 278:1872–1873, December 12,1997.

Leibeintz R. The NIH postdoc experience. *The Scientist*, p. 13, May 10, 1999.

Leyner M. How to avoid Salinger syndrome. *Time Magazine*, July 5, 1999.

May RM. The scientific investments of nations. *Science* 281:49–51, July 3, 1998.

Mezard M. Passing messages between disciplines. *Science* 301:1685–1686, September 19, 2003.

NIH grantees: Where have all the young ones gone? *Science* 298:40–41, October 4, 2002.

Peters T. The hottest jobs of the future. *Time Magazine*, May 22, 2000.

Phelan SE, Daniels MG and Hewitt L. The costs and benefits of clinical education. *Laboratory Medicine* 30(11):714–720, November 1999.

Seaman B. What makes a good college? *Time Magazine*, October 26, 1998.

Wallis C, Cole W, Mitchell E, Rutherford M and Tuff SJM. How to make your child a better student. *Time Magazine*, October 19, 1998.

Zielinska E. The scientist as politician. *The Scientist*, p. 73, September 2008.

Chapter 10

Reflections

Do Not be Afraid of Criticism

From the time we are children our actions are scrutinized — first by parents, then teachers, and eventually by our peers. Along with scrutiny comes criticism. For the most part, people criticize us to help us see a part of ourselves we are blind to, and even though sometimes it is incredibly painful to be made aware of those flaws and weaknesses, it is in our best interest to pay attention. When others critically examine our character, behavior, or work, our bodies react. If the feedback is positive, one might feel a surge of confidence, but if we don't like what we hear, our reactions might oscillate between defensiveness, panic, even rage. But what might at first seem like a personal affront can in the end lead to very important self-realizations. With patience we can learn to see that criticism, in its best form, is really highly personalized advice.

The scientific apprentice is not immune to criticism from mentors and peers, and, if well crafted and insightful, constructive criticism in the form of critiques can help advance his or her career, of which critiques are a continuous part of. Any publication submitted through a peer-review system receives a critique from the reviewers, as do grant applications (even successful ones). Even promotions involve a certain level of feedback about performance. We receive feedback about almost everything we do in the scientific research arena, and at a certain point not only do we become used to it, but we rely on it as a tool to help us see what we did wrong and what we could do better. When a manuscript is rejected, when grant applications score poorly, or if we are passed up for a promotion, our first reaction should be to ask for the critique.

How do We Take Critiques?

It is a very difficult thing to see, in detail, all the ways that your work is lacking in substance, logic or feasibility, and everybody reacts differently. Some people become extremely disturbed when they receive negative feedback, some so much that their friends and family avoid them until they have adjusted to the comments. Others enter a depressive mood that can last several days until the feedback is processed and the scientist is able to cope with criticism and face the problems point by point. I personally deal with critiques by not doing anything for the first twenty-four hours. I read them, put them aside, and go about my day without uttering so much as a word about them. The next day I start reviewing them and evaluating the comments one by one. The point is, I welcome criticism because it gives me a sense that people who are part of my environment still respond to my ideas in one way or another. Criticism is a healthy response to our activities and the only time we should be truly worried by it is when it stops, because that means no one is interested enough to tell you what they think so you can become better at what you do.

The scientific apprentice must learn to welcome criticism of all types. Even when it is offered with a bias, it provides us a tremendous amount of information regarding our scientific environment and our role in it. The scientific apprentice should learn to use critiques as a way to detect whether their work is relevant to the scientific environment in which they operate.

In my experience I have learned that the best way to eliminate someone from a position or a group, is to push them into isolation. Ceasing to provide feedback often does this, and without feedback, a person becomes transparent, invisible and their professional network is gradually closed off to them. No matter what they do they are unable to elicit a reaction from their superiors or other bodies within their institution. It is often the case that someone is greatly disliked — for a variety of reasons — yet they cannot be dismissed, either because there are no legitimate grounds for dismissal, or despite their ill repute they still draw significant funding to their Institution. Yet as they are squeezed out of the cadre, they recede into the background and their opinions aren't counted and they never receive feedback, mention, or critique. This image, while extreme, reinforces the point that the scientific apprentice must not be afraid of critical appraisal. On the contrary, they should welcome it and use it as reaffirmation of their relevance in the scientific community.

All scientists, as well as scientific apprentices, have had, or will have, at least one experience in which after delivering a lecture, instead of being bombarded with questions they are bombarded by a tremendous silence. No questions, no feedback, just the quick shuffling of bodies out of the room. Silence can send a powerful message. The absence of questions or criticism is very significant and needs to be carefully interpreted. First and foremost, do not blame the audience or the institution. Look to yourself first before you jump to a reactionary interpretation, such as believing that the audience didn't have the proper background to understand your talk, or that a mediocre institution does not value high-ranked individuals such as yourself, or that the reason for the lack of feedback is because you are so ahead of your time no one is equipped to critically evaluate your work. While any of these explanations might be accurate, do not consider them until you've conducted a thorough analysis of yourself as a scientific apprentice or scientist. In addition, I strongly suggest discussing the situation with a friend or with somebody you respect.

Are All Critiques Justified?

As important as they are, in reality, some critiques are not good critiques, not because the listener doesn't like what they say about his or her work, but because they are not well formulated or presented. Poor critiques often originate as an overreaction on the part of a reviewer who either lacks knowledge on the subject in question, sustains a personal bias, or simply because the reviewer has not taken enough time or care to understand the manuscript or application. If this kind of reaction is expressed during a conference, the scientific apprentice can politely approach the person in question. I have found that most of the time direct contact softens and clarifies an overly harsh critique. Questions that may arise from the critiques of a grant application can be addressed to the secretary of the study session that reviewed the application. In some cases this person can elaborate the meaning of the comments so weak areas can be better addressed in the introductory part of the grant resubmission. Lastly, the critiques of a paper submitted for publication are the most challenging. If upon considering the reviewers' critiques the editor considers the flaws grave enough they will reject the paper. In this case there is no way to answer or address the questions posed; the important thing is to carefully consider the reviewer's comments before submitting to another journal. In all of these cases it is crucial to analyze each critique carefully, eschew any feeling of self-pity, and become a disciplined critic of yourself. There is often a good chance that some truth and wisdom lays within each and every critique.

Why are Critiques so Important?

The scientific apprentice's leitmotif is to develop a research idea, discover a new paradigm, and make a mark in the history of science. For this reason the scientific apprentice needs, as does every scientist regardless of experience, critical evaluation of their work. Critiques are like multi-faceted mirrors that allow one to see how their work is perceived by their peers from different angles. This is not to say that I particularly look forward to having the work I have labored over picked apart flaw by flaw, but I welcome them for that very reason. We all need to understand the value of constructive criticism and learn early on in life to value it. *Criticism helps us evaluate the meaning of the world around us, the values of others, and teaches us not only how to handle the fracturing of our egos, but how to critique others as well.* It is important to be part of this exercise of being evaluated and to evaluate others because that is what leads a scientist to wisdom. Once the scientific apprentice has mastered the ability to be a good critic, their opinions are requested and welcomed.

Do Critiques Have a Long-Term Effect on the Scientific Apprentice?

There is no doubt, for the reasons explained above, that the exercise of being under the spotlight of peer-reviewed critiques has short and long-term impacts on the scientific apprentice. The short-term effect is apprehension, which is natural because a critique that impinges on the data collected or on the conclusions reached can cause a significant delay or even result in the need for a whole set of experiments to be repeated. Therefore, it is the function of the mentor to coach the scientific apprentice in a way that critiques take place on a regular, even daily, basis so as not to allow the scientific apprentice to accumulate a massive amount of mistakes that will require a heavy set of changes and create a more difficult and frustrating situation to respond to. Because the nature of scientific critiques will be new to most scientific apprentices, the mentor is also tasked with training the apprentice to receive them so that they become a natural part of the working process.

The long-term effect of critiques is that, over time, the scientific apprentice will appreciate how much the feedback of others has helped his or her career and personal development as a scientist.

What Do Critiques Say About the Mentor?

The mentor definitely has a central role in training the scientific apprentice on how to receive critiques and eventually how to formulate them. If the mentor lives in a spirit of greatness, whatever he or she does will have a positive influence on the scientific apprentice's life in science. If the mentoring scientist, however, is learning the trade at the same pace as the scientific apprentice, both will suffer from inexperience. I mentioned a moment ago *"greatness,"* a word we commonly identify with largeness, magnificence, and power. While greatness is all these things, it can also signify the scale of emotion or feeling. Greatness in science, therefore, means that everything we do is full of meaning; that we participate in scientific thought because we believe in its inherent power and majesty; that we want to make a better world and in turn better ourselves; that what we do what needs to be done for the good of humanity. *To be a great person is to separate action and thoughts from pettiness and selfishness, and in the process become more in tune with the world. A great mentor understands that his or her role as a teacher is to open the door to that world.*

The Workplace

The time will inevitably arrive for the scientific apprentice to find a position in their area of interest, or more accurately, a place where their full potential can develop and unfold. This search does not end with the scientific apprentice, but is constant for all of us, for we should never stop searching for the opportunity to be the best of ourselves.

The scientific apprentice should ask themselves not *where* they want to work, but what kind of research they want to spend their life doing. Perhaps one scientific apprentice is inclined towards academic work in a university, while another longs to do research at a government institution, or a pharmaceutical company. Others might want to be in a place where research is all they do, all the time. All of these are great options, and it is not the objective of this book to provide guidelines for selecting an institution, or establishing pros and cons for each. What I will provide, however, in keeping with the spirit of this book, are certain questions that the scientific apprentice needs to address, either through observation, by direct inquiry, or by simply listening carefully to what people have to say.

Is the Facility Well Maintained?

This question might seem minor in light of more pressing ones, but it is not. Management and cleanliness might not be important for a motivated scientific apprentice eager to work, but it is important to remember that buildings, labs and other physical elements of a place are a reflection of its management. Many years ago I was invited for an interview at a well-recognized east coast university. The prearrangements for my arrival were well executed — I was given clear directions and greeted warmly. Upon arriving at the university however, I was puzzled by the condition of the parking lot and the main entrance. The parking posts were broken, and most of the signs had been vandalized, making it difficult to read the number on my parking spot. The main lobby was clean and a very kind person was there waiting to take me to my interview. While we waited for an elevator in the lobby, my host apologized for the delay, explaining that three of the four elevators were out of order and had been for quite some time. These details have nothing to do with the position they wanted to offer me, but I got a very strong feeling that the place was not in good economic standing. In the end I turned down the post, only to learn several months later that the university was admitting to significant economic stress. Aesthetic order is a sign of consideration for the people that work in a place, as well as a sign of how the institution is run. It also tells a lot about the general morale — if nobody cares what happens to the façade, chances are that the internal components are also poorly tended to.

Are the Laboratories Well Occupied or Simply Vast Spaces Dotted With a Few Human Beings?

It is true that the external appearance of the parking lot or the lobby of the hosting institution can deceive you. But what about the laboratory space — what can the lab itself tell us about what it's like to work there? First the scientific apprentice needs to have an idea of how to estimate space. This becomes more important later on when writing grants and proposals because space becomes a factor in terms of efficiency. It is good to begin looking at the big picture from the start. Because you can hardly pull out a measuring tape during an interview, use your living quarters to start developing your sense of spatial size. In general, a small bedroom is between 100 and 150 sq. ft. A master bedroom could be anywhere between 100 to 600 sq. ft. With the bedroom as a reference, get a sense in your mind of your physical relationship to 100 sq. ft. With this dimension in mind, if once you arrive at a lab you notice that there is one postdoctoral or technical person every 100 sq. ft, from this it can easily be concluded that the lab is a busy place and that the space is well utilized. If, however, you see one person occupying a lab of 1000 sq. ft., then you might start to suspect that the number of inhabitants in that scientific community is not very large. Whenever I visit large universities in the United States and Europe (but less in Latin America), there are many large, spacious, shiny laboratories up to 2000 sq. ft. occupied by only one of two technical or postdoctoral people. Upon visiting a large pharmaceutical company outside Philadelphia I noticed a huge laboratory upwards of

1500 sq. ft. was occupied by a single person. It might be the case that a vast laboratory full of instruments and modern equipment is staffed by a person so capable he's able to use them all at once. It could also be that my perception of one person per 100 sq. ft. as a good ratio is flawed for the same reason. There is nothing to say that a few well-selected people cannot achieve a lot. But if an ill populated lab is also empty in regards to equipment and supplies, or it looks like the benches have not been used in a while, this could mean that there is little work going on or that the space and resources are poorly distributed. But let's say you walk into a laboratory and you see ten busy people working with a sense of concentrated urgency, equipment running and benches full of working material. You see another scientific apprentice discussing data with the mentor and learn that the leader of the group is working on a grant or a paper and nobody in the lab is distracted by a visitor's presence. This is the kind of laboratory that speaks for itself and is likely to be part of an active and supportive institution.

Yet one laboratory could still be misleading. But if you visit a facility — for example a genomic facility — and see that within a mid-sized space are the most modern array equipment with five people working diligently and mining data, this too, is a good sign. If the scientific apprentice wants to get an accurate picture they should ask to visit other laboratories and see if the same phenomena takes place elsewhere. All of these are parameters that indicate the type of place that the scientific apprentice would like to consider.

In visiting the laboratories, survey the type of equipment available. Are they new? Are they operational or are covered in plastic because nobody uses them? Observe other scientific apprentice's desks — are they free of papers or any indication of active work? Do you see them working? Everything you see is an indication of a lab's pulse.

Additionally, observe the principal investigator's office. Is it clean but devoid of any indication of active work, or can you perceive of the signs of an active intellectual life? Is the principal investigator listening to you, attentive to what you are saying and interested in what you want to do, or watching the clock to see when the interview will be over? This is important because the type of principal investigator prevalent in an institution indicates its quality.

What is the Cultural Composition of the Lab?

The scientific apprentice has assessed the physical components of the potential workplace but now is the time to analyze their prospective colleagues. Is this a place where different cultural groups converge, or is it a mono-ethnic group in which all members share a common background and experience? This is important because not all the scientific apprentices can assimilate very well to other cultures or be comfortable as a minority in a predominant group. Due to the nature of the United States, however, it is likely that any given lab will be made up of people from a variety of backgrounds so it behooves the scientific apprentice to expand their familiarity with foreign cultures. While this is by no means the most important part of the first impression, it is worth making the observation, especially to see how people interact in an academic context. If the scientific apprentice is asked to give a seminar, this is the best scenario in which to get a sense of prospective colleagues, not only during the presentation but afterward when the question period starts. The real challenge is to determine the intellectual and scholarly level of the place. The questions asked after the lecture will reveal a lot. If questions are mainly coming from the senior members and not from the scientific apprentices, this indicates that the young investigators are not involved or are not curious enough. Are the questions relevant or perfunctory?

The Offer

Lastly, the scientific apprentice will need to meet the members of the institution that make the final decision and the final offer. If the scientific apprentice feels that this is the place for them, this final interview might be anxiety ridden, but surely not the most difficult part of the process. By this point, decisions have been made by both parties and all the scientific apprentice needs to do is listen, and if offered the job, give thanks and express pleasure and eagerness about becoming part of the group. If the institution decides not to hire the scientific apprentice, there is nothing the scientific apprentice can do at this point to change their mind. All they can do is be gracious, thank them for their consideration and wish them success in finding the right candidate.

All this might sound obvious or redundant but it is not. The decision to hire does not take place in the last interview but begins in the initial process and continues during the scientific apprentice's entire visit. Just as the scientific apprentice has been analyzing the institution, the institutions are making the same evaluations of the scientific apprentice.

If the scientific apprentice is rejected in this first round of job hunting they must remember that not all negative experience turn out to be bad. I am grateful that many things that I had once wanted so badly did not come to be. Learn from the experience and stay ready for the next search. *After all, you are an apprentice of science and even the most seasoned scientist must never forget that until the end of our last useful hour of existence we must keep alive that same spirit of learning that moved us towards science in the first place; the spirit that ensures that we will forever be apprentices of science.*

Suggested Readings

Amador PAG. en el mundo de la medicina. *Medico Interamericano* p. 49.

Bignami GF. The microscope's coat of arms. *Nature* **405**:999, June 29, 2000.

Bloom H. Magic words. Time bonus sections generations. G10. July 2002.

Brooke J. Science and religion. Lessons from history? *Science* **282**:1985–1986, December 11, 1998.

Bunk S. Into the future. September 22, 2003.

Janeway CA, Jr. *Nature* **423**:237, May 15, 2003.

Check E. Harmful potential of viral vectors fuels doubts over gene therapy. *Nature* **423**:573–574, June 5, 2003.

Daenke S. Courage could win back confidence in science. *Nature* **401**:321, September 23, 1999.

Davis EB. A God who does not itemize versus a science of the sacred. *American Scientist* **86**:572–574.

Declines in NIH research grant funding. *Science* **322**:189, October 10, 2008.

Flower R. The man who knew doses. *Nature* **406**:831, August 24, 2000.

Friedrich MJ. Profiles. *Laboratory Medicine* **29**:719–720, November 1998.

Gazzangia MS. How to change the university. *Science* 282:237, October 9, 1998.

Gibbs N. It's only me. *Time Magazine*, March 19, 2001.

Giles J. The man they love to hate. *Nature* 423:216–218, May 15, 2003.

How head-hunters track down the winners for science's top jobs. *Nature* 401:407–409, September 23, 1999.

Howell KJ. The Great Dane of Urinaborg. *Nature* 405:883–884, June 22, 2000.

Hunter M. Work, work, work. *Modern Maturity*, pp. 37–49, May–June 1999.

Igo SE. Bending behavior. *American Scientist*, pp. 267–268, May–June 2006.

Ihde D. Epistemology engines. *Nature* 406, p. 21, July 6, 2000.

Jay V. The legacy of Reinier de Graaf. *Arch Pathol Lab Med* 124:1115–1116, August 2000.

King M-C and Motulsky AG. Mapping human history. *Science* 298:2342–2343, December 20, 2002.

Maienschein J. Scientific literacy. *Science* 281:917, August 14, 1998.

Malakoff D. Bayes offers a "new" way to make sense of numbers. *Science* 286:1460–1464, November 19, 1999.

Malartre E. Words, words, words. *Nature* 406:833, August 24, 2000.

Marshall E. Lemons, oranges and complexity. *Science* 322:209, October 10, 2008.

Mermin ND. Varieties of silly experience. *Nature* 407:17–18, September 7, 2000.

Normile D. Clinical trials guidelines at odds with U.S. policy. *Science* 322:516, October 24, 2008.

Nossal GJV. A troubled pilgrim's progress. *Nature* 424:253–254, July 17, 2003.

Pincock S. A planck walk. *The Scientist*, p. 46, July 2008.

Schwartz J. Making genetic history. *Nature* 453:1181–1182, June 26, 2008.

Shaulsky G. The cheatin' amoeba. *The Scientist*, p. 30, July 2008.

Vogel S. Academically correct biological science. *American Scientist* 86:504–506.

Westheimer G. How we see things. *Science* 288:2324, June 30, 2000.

While the lab coats gather dust. *Science* 301:593, August 1, 2003.

Wolpert L. The well-spring. *Nature* 405:887, June 22, 2000.

Index